Anatomy and Physiology for Nurses

For Elsevier

Senior Commissioning Editor: Ninette Premdas
Associate Editor: Dinah Thom
Project Manager: Caroline Horton
Designer: Judith Wright
Illustration Manager: Bruce Hogarth

Anatomy and Physiology for Nurses

Roger Watson BSc PhD RGN CBiol FIBiol ILTM FRSA

Professor of Nursing, Graduate School of Nursing,
University of Sheffield, Sheffield, UK

Illustrations by **Amanda Williams**

TWELFTH EDITION

ELSEVIER

EDINBURGH LONDON NEW YORK OXFORD PHILADELPHIA ST LOUIS SYDNEY TORONTO 2005

ELSEVIER

First edition 1939
Ninth edition 1979
Tenth edition 1995, all previous editions were published in the Nurses' Aids Series
Eleventh edition 2000
Twelfth edition 2005
    Reprint 2006, 2007

ISBN-13: 978–0–7020–2749–9
ISBN-10: 0–7020–2749–9

British Library Cataloguing in Publication Data
A catalogue record for this book is available from the British Library

Library of Congress Cataloging in Publication Data
A catalogue record for this book is available from the Library of Congress

Notice
Knowledge and best practice in this field are constantly changing. As new research and experience broaden our knowledge, changes in practice, treatment and drug therapy may become necessary or appropriate. Readers are advised to check the most current information provided (i) on procedures featured or (ii) by the manufacturer of each product to be administered, to verify the recommended dose or formula, the method and duration of administration, and contraindications. It is the responsibility of the practitioner, relying on experience and knowledge of the patient, to make diagnoses, to determine dosages and the best treatment for each individual patient, and to take all appropriate safety precautions. To the fullest extent of the law, neither the publisher nor the authors assume any liability for any injury and/or damage.

The Publisher

ELSEVIER your source for books, journals and multimedia in the health sciences
www.elsevierhealth.com

Working together to grow
libraries in developing countries
www.elsevier.com | www.bookaid.org | www.sabre.org

ELSEVIER   BOOK AID International   Sabre Foundation

The publisher's policy is to use **paper manufactured from sustainable forests**

Printed in China

Contents

SECTION 5 NUTRITION AND ELIMINATION 277

SECTION 6 PROTECTION AND REPRODUCTION 345

Preface

The twelfth edition of *Anatomy and Physiology for Nurses* retains what is best in previous editions with the careful introduction of some new material. I am keen that the book remains useful to those nurses, and other health professionals, who need a clear and comprehensive introduction to anatomy and physiology. To achieve this I have tried to present the book in a style that continues to be easy to read and which requires a minimum of reference to illustrations, tables or appendices. The main new feature of this book is in the illustrations, which have all been redrawn. There has been minimal reorganization of the contents, and the bibliography – in this age of the World Wide Web – has been shortened.

I am grateful to reviewers of the eleventh edition who suggested many of the changes that are apparent in the present edition, especially the addition of genetics. I remain grateful to the publishers for their continuing support for this small volume, and hope that readers find it useful and that it opens up for them other more advanced books on the subject.

Finally, my wife and family deserve a special mention for tolerating my long-standing relationship with the word processor.

Roger Watson
Hull, 2005

SECTION 1

The characteristics of living matter

SECTION CONTENTS

SECTION INTRODUCTION

The first section of this text includes material which is fundamental to an understanding of subsequent sections. The material is presented in a logical way, beginning with the simplest level of organization of matter up to the most complex level found in the human body, of tissues, organs and systems.

The chemical level of organization considers the strong bonds which occur between atoms in the formation of molecules; these are the covalent bonds. Such bonds between carbon and hydrogen are the basis of the building of hydrocarbon molecules, which represent the basic structure of all biological macromolecules – the lipids, carbohydrates, proteins and nucleic acids. Weak interactions, principally the hydrogen bond, which is an essential feature of the structure and function of proteins and nucleic acids, are also considered.

At the cellular level of organization the plasma membrane, which contains the cellular contents and also controls the intracellular environment, is considered. The transport processes which work to exchange molecules and water between the extracellular and intracellular environments are described and the processes whereby cells divide to produce either identical daughter cells or gametes for subsequent involvement in sexual reproduction are introduced.

Finally, the highest levels of organization, which involve the organization of cells into tissues, tissues into organs and organs into systems, are presented. Before leaving this section the concept of homeostasis, which maintains the internal environment of the organisms, is introduced. Homeostasis is a unifying concept in physiology as it involves all of the systems working together to respond to changing circumstances and stresses imposed upon the organism by changes in the external environment or, in some cases, by disease.

Chapter 1

Introduction

LEARNING OBJECTIVES

After reading this chapter you should understand:

- the physical states and changes in state of matter
- elements, compounds and mixtures
- acids and alkalis
- measurement in moles
- diffusion and osmosis
- the characteristics of living matter
- the terms used in anatomy

Nurses care for human beings in illness and during vulnerable periods of their lives. In order to do this properly they need to understand the structure and function of the body and to be able to relate this knowledge to the care of the patient. Each organ in the body plays its part in maintaining the health of the whole, and if one organ is at fault the whole body will be affected. The structure of each part suggests the function, and the function suggests the structure, so that the study of the human body is a logical process of thinking and reasoning, not merely of memorizing.

Terms

1. *Anatomy* is the study of the structure of the body.
2. *Physiology* is the study of its function.

Some knowledge of elementary physics and chemistry will help in the study of anatomy and physiology and will also be needed to enable the nurse to care for the patient and to carry out nursing procedures.

1. *Chemistry* deals with the composition of matter and the reactions between various types of matter.
2. *Physics* deals with the behaviour and characteristics of matter, for example, whether it gives off heat and light, or conducts electricity.

MATTER

Matter is anything that can occupy space. When we refer to a 'space' we usually mean that it contains air, which can be displaced by other forms of matter. For example, an 'empty' bottle is not really empty as it contains air. A space from which the air has been extracted is called a *vacuum*.

Physical states of matter

Matter has three physical states: solid; liquid; and gas. A solid does not easily alter in either shape or size, e.g. stone, brick.

A liquid takes the shape of the vessel it is put into but does not alter in size, e.g. 500 millilitres (mL) of water has no shape but it takes the shape of the container it is put into; if it is put into a jug which will hold 1 litre, the jug will not be filled.

A gas (e.g. oxygen, hydrogen or nitrogen) takes the shape and size of the vessel containing it. If the air is extracted from a container and a small quantity of gas is introduced the gas will expand to fill the container. If more gas is introduced the gas in the container is compressed, and the small particles which make up the gas are packed more closely together. More gas could be introduced until the flask bursts. A common example of this can be seen when a bicycle tyre is pumped up. After a few pumps the tyre seems full but when someone gets on the bicycle the tyre goes flat because there is so little air in it. You can go on pumping until there is so much air in the tyre that when you get on the bicycle your weight does not flatten the tyre. A measured amount of gas in a cylinder will exert a known pressure. As the gas is used up the pressure in the cylinder will drop and it is in this way that the quantity of gas remaining in the cylinder can be gauged.

Change of state

At normal temperatures, iron is solid and water is liquid, but the state of matter can alter. The changes are due to heating or loss of heat. For example, liquid water becomes solid ice when heat is lost, and solid butter becomes liquid oil when it is warmed. The changes are described as:

- Melting – the change of a solid into a liquid as a result of heating, e.g. the turning of ice into water
- Evaporation – the change of a liquid into a gas as a result of heating, e.g. the turning of water into steam or water vapour. Gases formed by evaporation are called vapours
- Condensation – the turning of a gas into a liquid as a result of cooling, e.g. the turning of water vapour into water
- Consolidation – the turning of a liquid into a solid as a result of cooling, e.g. the turning of water into ice.

The first two of these changes, melting and evaporation, are due to heating, but the heat which causes this change of state does not cause any rise in the temperature of the matter. Heat is a form of energy and the energy in this case is used up in producing the change of state, so it does not heat the matter. For example, the temperature of melting ice is 0°C. When the ice melts the water is also at a temperature of 0°C. Water boils at 100°C. If you leave the kettle on the heat, the water will not get any hotter because the added heat is used up in turning the water into water vapour. In the same way, heat from the body evaporates the sweat from the skin, excess body heat is used up and the body is cooled. Heat used up in this way is called latent heat. In condensation or consolidation, latent heat is set free.

ELEMENTS, COMPOUNDS AND MIXTURES

Some substances cannot be split up and these substances are called elements. Over 90 elements are known to exist, including oxygen, carbon, nitrogen, iron, silver and gold. Substances which can be split into different elements are of two types: compounds and mixtures.

A compound is a substance made of two or more elements which combine chemically to form a new substance with new properties. For example, hydrogen is a very light gas used to fill balloons; it is highly inflammable and may explode. Oxygen is a gas which supports combustion but will not itself burn. These two gases in combination make a compound, water, which is a fluid, but it is not inflammable, will not explode, will not support combustion and will not burn. In fact it is very useful for putting out a fire. The elements that form a compound are always present in fixed proportions. For example, water consists of two parts of hydrogen combined with one part of oxygen. If the proportions were altered a compound might be formed but it would not be water. Two parts of hydrogen combined with two parts of oxygen give a compound called hydrogen peroxide which looks like water but has very different properties. Another characteristic of a compound is that it is not easy to separate the elements which compose it. To accomplish this separation a chemical change must take place.

A mixture is a substance made of two or more elements mixed together but not chemically combined. Air is a mixture of gases, mainly nitrogen and oxygen, with traces of carbon dioxide and other gases. A mixture has no new properties but simply those present in the elements which form the mixture. Air supports combustion because it contains free oxygen but it does not support combustion as well as pure oxygen because of the high percentage of nitrogen, which will not support combustion. The elements in a mixture can be separated fairly easily because they are not chemically combined. Oxygen in air will support combustion because it is 'free' oxygen. Oxygen in water is not free and therefore will not support combustion. In a mixture there is no fixed proportion of elements composing it. Three different samples of air may contain three different proportions of oxygen, yet each is a specimen of air.

GAS PRESSURE

The molecules of a solid are strongly attracted to one another, so a solid does not easily alter in size or shape. The molecules of a liquid are less strongly attracted to each other, so their positions can be altered more easily and the liquid takes the shape of the container. The molecules of a gas are attracted to each other very little so that when gas is put into a vacuum container the molecules separate from one another and fill the flask. If more gas is introduced the molecules become more densely packed and the gas is said to be under pressure. The pressure of gases is important in their use in the body. Unless oxygen is under sufficient pressure in the lungs it will not be transferred to the blood for transport round the body and the body tissues will suffer from lack of oxygen. On a high mountain the pressure of the air is so low that humans cannot live without an extra supply of oxygen, though a candle will burn because there is sufficient quantity of air even though the pressure is reduced. In an airless room where there is reduced quantity of oxygen the candle will not burn, though a person can continue to live as the oxygen present is still under pressure.

SOLUTIONS AND EMULSIONS

When a lump of sugar is put into a cup of hot water the sugar dissolves and a solution of sugar is present in the cup. The molecules of sugar mix with the molecules of water until the sugar is evenly distributed throughout the fluid. The process can be speeded up by stirring but the even distribution of sugar will occur eventually without stirring. A measured quantity of fluid will only dissolve a measured quantity of another substance, and when the fluid has dissolved as much as possible, it is called a *saturated solution*. Fluids can dissolve solids, liquids and gases. Oxygen is dissolved in water, which enables fish to live in it. Heating a fluid drives off the dissolved gas, which appears at the surface as bubbles as the fluid gets hot. Water is the most common solvent and can dissolve a wide variety of substances, but substances which cannot be dissolved in water may be dissolved in other liquids. For example, water will not dissolve oil, but oil will dissolve in methylated spirit. The fact that water is a common solvent makes it important in the body, two-thirds of which is water. Most of this water is in the cells which make up the body: some of it surrounds the cells, which can only live in a salt solution, called saline, and some is in the circulating blood and lymph.

An *emulsion* of oil is different from a solution. If oil is mixed with water the oil will rise to the top when stirring is stopped and will form a layer of pure oil on top of the water. If the oil is mixed with a solution of sodium carbonate (washing soda), an emulsion will be formed. The oil will be broken up into little droplets which will be suspended in the fluid and will not run together again. When allowed to stand the oil will still rise to the top, but the fat droplets will remain separated. Milk is a natural emulsion and the droplets of fat which form the cream can easily be seen under a microscope.

ACIDS AND ALKALIS

Substances may be either acid or alkaline in reaction, or they may be neutral (neither acid nor alkaline). The simplest agent used to detect the reaction of a substance is *litmus* which can be obtained as a fluid but is conveniently used in the form of litmus paper. Blue litmus paper will turn pink when in contact with acid. Pink litmus paper will turn blue when in contact with an alkali. Neutral substances turn both pink and blue litmus paper a purple colour. Every living cell must have the correct level of acidity or alkalinity if it is to survive. The blood and body tissues are slightly alkaline and vary little throughout life. If too much acid or too much alkali is present in the body, illness will occur, and if there is an alteration in the reaction of the blood or body tissues, death will follow. It is necessary to know the degree of acidity or alkalinity of certain substances and for this purpose, pH paper is used. It changes colour to indicate the pH of the substance being tested and the colour is checked against a chart which is specific to the range of the pH paper. The pH scale runs from pH1 for the strong acid to pH14 for the strong alkali. Neutral is pH7 and the pH of blood is 7.4.

Pure water is neutral in reaction and as previously stated, consists of molecules composed of two atoms of hydrogen and one atom of oxygen. Some of the molecules are broken up into smaller particles, e.g. hydrogen ions (H ions) or hydroxyl ions (OH ions). H ions have a positive electrical charge because the atoms have lost an electron and OH ions a negative electrical charge. In pure water the H ions and the OH ions are equal in number and the reaction is therefore neutral. In some solutions, H ions may outnumber OH ions and the reaction of the solution will therefore be acid. If the OH ions outnumber the H ions the reaction will be alkaline. The pH shows the concentration of hydrogen ions (the H ion concentration), strong acids, e.g. hydrochloric acid, having a high H ion concentration, and strong alkalis, e.g. caustic soda, the highest concentration of hydroxyl ions (OH ions). Some substances, which slow down changes in pH, are called buffer substances. In an alkaline solution such as body fluids, buffer substances neutralize acids or any excess of alkalis which may be added to it or produced in it. Examples of buffer substances are sodium, potassium and protein, which serve as buffers against acid substances in the body, and carbonic acid, lactic acid and fatty acids, which serve as buffers against any excess of alkaline bases.

MINERAL SALTS

A mineral salt is a substance made by the action of acid on a mineral. The mineral is called the base of the salt, and the scientific name shows the base and the acid which together form the salt, e.g. sodium chloride is made by the action of hydrochloric acid on sodium. A salt may be acid, alkaline or neutral in reaction. A strong acid acting on a strong alkali forms a neutral salt; a strong acid acting on a weak alkali forms an acid salt; a weak acid acting on a strong alkali forms an alkaline salt. The various salts present in the body differ only slightly in reaction.

MEASURING THE AMOUNT OF A SUBSTANCE

The number of neutrons added to the number of protons in an atom gives its atomic weight. Hydrogen has an atomic weight of one (as it has one proton and no neutrons) and oxygen an atomic weight of 16 (as it has eight protons and eight neutrons). A molecule of water, which consists of two hydrogen atoms and one oxygen atom (H_2O), will therefore have a weight of 18. This is known as its *molecular weight*. Understanding atomic and molecular weights is important in understanding how the amount of a substance is measured. The concept of atomic weight will be explained in more detail in Chapter 2.

The amount of a substance can be expressed in terms of its mass, but in physiology and biochemistry it is more important to know *how many particles* (atomic, ionic or molecular) of the substance are present rather than their total mass. However, since even tiny amounts of a substance will contain many millions of particles, it would be very cumbersome to use such huge numbers. Therefore a unit defined as a standard large number of particles was devised. The number of particles chosen was the number of particles in that amount of a substance that has a mass equal to its molecular or atomic weight in grams; this number will always be 6.023×10^{23} and is referred to as Avogadro's number. For example, the atomic weight of carbon is 12, therefore 12 g of carbon will contain 6.023×10^{23} atoms of carbon. The molecular weight of water is 18, therefore 18 g of water will also contain 6.023×10^{23} molecules of water. The amount of substance containing this number of particles is called a *mole*, the symbol for which is mol.

The mole can be used to indicate the concentration of a substance in a solution. If one mole of a substance is added to one litre of water, the concentration of the substance in the resulting solution will be 1 mol/litre. Because the mole is a measure of the number of particles, one mole of a molecule that breaks up into ions when added to water will produce one mole of each of the ions produced. For example, hydrochloric acid (HCl) breaks up into hydrogen ions (H^+) and chloride ions (Cl^-) when added to water. One molecule of HCl will produce one H^+ and one Cl^-. Therefore 6.023×10^{23} molecules of HCl (i.e. one mole of HCl) will produce 6.023×10^{23} H^+ (one mole of H^+) and $6.023 \times 10^{23} Cl^-$ (one mole of Cl^-).

The mole is often too large a unit to describe the concentrations of substances encountered in clinical medicine. Therefore a prefix is used, as for other units. The prefix commonly used is milli-, one millimole (mmol) being 1000 times smaller than a mole. For example, the concentration of sodium chloride (NaCl) in the blood is approximately 0.15 mol/L but this is usually written as 150 mmol/L. A solution made up to have this same concentration of NaCl (i.e. 150 mmol/L) is known as *isotonic* or physiological saline.

Moles cannot be used for measuring the concentration of protein substances because the molecules are so complex that their molecular weight cannot be determined accurately. Such substances are therefore expressed in terms of their mass, i.e. in grams per litre or in milligrams (mg) per 100 mL. It is rarely necessary to convert from one to the other but if it should be the formula is:

$$\text{Concentration in mmol/L} = (\text{Concentration in mg/100 mL} \times 10)/\text{Molecular weight of substance}$$

The concentrations of dissolved gases, e.g. in the blood, are often stated in terms of pressure rather than moles or grams because pressures are easier to measure.

DIFFUSION

If two gases of different composition come into contact, intermingling of the gases takes place until the composition of both is the

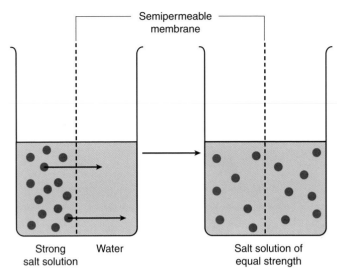

Fig. 1.1   Diagram to show the diffusion of liquids.

same. For example, the air we breathe out contains more carbon dioxide and less oxygen than the air we breathe in, but the expired air does not float around the room as a portion of different air. The air in the whole room will eventually contain less oxygen and more carbon dioxide overall. This process is known as *diffusion* (Fig. 1.1). Diffusion is also used to describe the passage of small molecules of acids and salts through a semipermeable membrane. A membrane may be described as permeable when it allows substances in solution to pass through, as impermeable when no fluid can pass through, and as semipermeable when water and small molecules in solution can pass through but not larger molecules. If a strong solution of salt is separated from a weaker solution by a semipermeable membrane the salt molecules will pass through the membrane from the stronger solution to the weaker one until both solutions are of equal strength.

OSMOSIS

If a substance with large molecules, e.g. sugar, is made into a solution and is separated from a weaker sugar solution by a semipermeable membrane, only water will pass through the membrane, from the weaker solution to the stronger solution, because the sugar

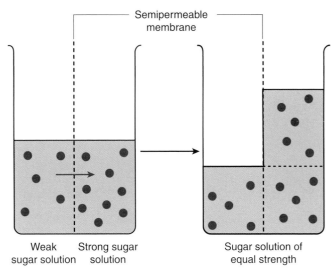

Fig. 1.2 Diagram to show osmosis.

molecules are too large to pass through. This passage of water across a semipermeable membrane is called osmosis (Fig. 1.2).

SPECIFIC GRAVITY

Specific gravity is the weight of a known volume of liquid divided by the weight of an equal volume of pure water.

Specific gravity = Weight of substance/Weight of equal volume of pure water

The specific gravity of pure water is expressed as 1.000. The specific gravity of urine ranges from 1.010 to 1.020, and that of blood is about 1.055.

ORGANIC MATTER AND INORGANIC MATTER

Organic matter is that which is living or has been alive, e.g. wood or coal. Inorganic matter is that which is not living and never has been alive, e.g. water or iron. Inorganic substances can be built up into organic compounds by living organisms.

All living matter is made up of small units called *cells*. Some organisms, such as bacteria, consist of one single cell; others, like human beings, consist of many hundreds of millions of cells, all functioning together to make a complete whole. All living cells, however simple, have certain characteristics which are always present. These characteristics are:

- Activity
- Respiration
- Digestion and absorption of food
- Excretion
- Growth and repair
- Reproduction
- Irritability.

Activity

This is the most striking characteristic of living matter. It is more apparent in the animal world than in the vegetable world, as the animal must move about in search of food, but even in plants the buds can be seen to break out in the spring, and with a microscope, activity becomes as obvious in the plant as in the animal. There can never be any activity without energy. Cars are driven by energy derived from petrol, trains are generally moved by electrical energy, and in living matter the energy is also obtained by burning fuel. The fuel for the human body is the food eaten, particularly carbohydrates and fats. Oxygen is also necessary for the combustion of fuel, and living things obtain their oxygen from the air or water in which they live. Combustion of fuel also produces waste products, such as carbon dioxide and water, which must be disposed of.

The combustion of fuel in living matter produces some energy for work and some energy in the form of heat. The body is economical in that it produces more work energy and less heat for each unit of food consumed, but the heat produced by combustion is not altogether wasted for some heat is needed. The body can remain healthy only within a narrow temperature range of 36–37.5°C and any heat produced in excess of this must be eliminated if life is to continue.

Respiration

All living matter requires oxygen and gives off carbon dioxide. The oxygen is required for combustion, or oxidation, of food, and carbon dioxide is the waste product of combustion. This process of taking in oxygen and giving off carbon dioxide is called respiration and it is continuous throughout life. The amount of oxygen required and the quantity of carbon dioxide given off vary with the amount of activity taking place.

During sleep the human body requires comparatively little oxygen but it requires much more during strenuous exercise.

Digestion and absorption of food

All living matter requires food. Some food can be absorbed as it is but most of it must be broken down into substances which have smaller and more simple molecules before it can be absorbed. The process of breaking down complex foods into smaller substances is termed *digestion*. It is brought about by enzymes, which are themselves protein substances and which act upon the food, preparing it for *absorption*.

Excretion

All living matter produces waste products which are of no further use and must be removed. If allowed to accumulate, waste products will interfere with the processes of life. In the human body these waste products include carbon dioxide from the lungs, sweat from the skin, and urine from the kidneys.

Growth and repair

Living matter is able to build up new living matter, similar to itself, from food which is taken in. Protein foods include meat, cheese and milk, and these provide the building material for the body. Because the body is continually active, its parts are constantly being worn out and they are made good by the building of new living matter. In childhood, growth and repair are taking place simultaneously. In old age the building up does not keep pace with the breaking down, so weakness begins to appear and will eventually cause death unless illness supervenes. It is this ability to grow and to repair its own tissues, and also the power of reproduction, that makes living matter different from manmade substances.

Reproduction

All living matter can reproduce its own kind. In the simplest forms of life, reproduction is a very simple process, involving the splitting of the parent cell into two. In animals, female cells called ova are produced in the ovaries, and sperms, or male cells, are produced in the testes. The sperm cell must fertilize the ovum before reproduction can take place.

Irritability

All living matter is able to respond to a stimulus. In higher animals this means the power to be aware of the environment and to respond to it. Irritability is more marked in animals but it can be observed in plants. If a bulb is planted upside down the roots will still grow deep into the soil and the leaves into the air, or if a plant is placed in a dark corner it will grow tall and thin as it seeks the light. This indicates that even plants are aware of the environment. In animals, irritability allows the awareness of danger, the recognition of that which is useful, e.g. food and water, and the exercise of free will.

All this is what is meant when the physiologist talks about 'life'. Living matter is that which has all the characteristics mentioned above, and all these characteristics are the results of chemical changes. All living matter has the ability to produce protein substances called *enzymes* which in turn control and produce chemical changes within the cell. Enzymes cause other substances to combine with one another, or to split up, though they themselves do not enter into the reaction. For example, complex starches combine with water and split up into simple sugars in digestion; in metabolism, sugar splits up into carbon dioxide and water. All these changes are brought about by enzymes.

DEFINITION OF TERMS USED IN ANATOMY

In order to achieve uniformity of description, an *anatomical position* has been chosen and defined. The body is erect, facing the observer, with the arms at the sides and the hands facing forwards (Figs 1.3 and 1.4). Terms that are commonly used are:

- Superior: upper or above
- Inferior: lower or below
- Anterior or ventral: towards the front
- Posterior or dorsal: towards the back
- Distal: furthest from the source
- Proximal: nearest to the source

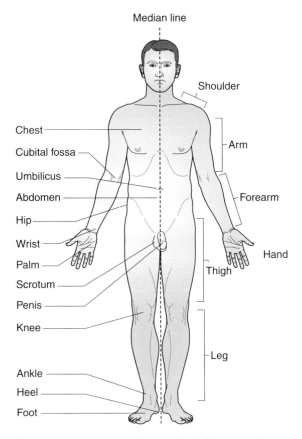

Fig. 1.3 The anatomical parts of the male body (front view).

- External: outer
- Internal: inner.

The *median or sagittal line* is an imaginary line passing vertically through the midline of the body from the crown of the head to the ground between the feet, dividing it into right and left halves. Lateral means furthest from the median line; medial means nearest to the median line.

1. A *horizontal* section divides the body into superior and inferior portions.
2. A *sagittal* section divides the body into right and left portions parallel to the median line.

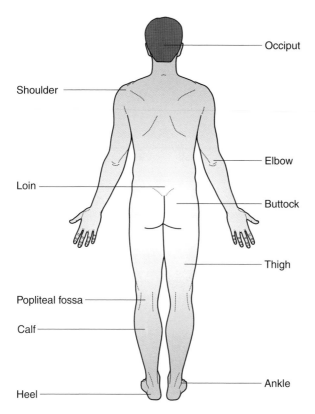

Fig. 1.4 The anatomical parts of the body (back view).

3. A *coronal* section divides the body into anterior and posterior portions.

Cavities of the body

The body has two major cavities, each subdivided into lesser cavities (Fig. 1.5).

The *ventral cavity*, within the trunk, is subdivided into:

1. The thorax or thoracic cavity
2. The abdomen, or abdominal cavity, which is continuous with the pelvic cavity.

The *dorsal cavity* is subdivided into:

1. The cranial portion, containing the brain
2. The spinal portion, containing the spinal cord.

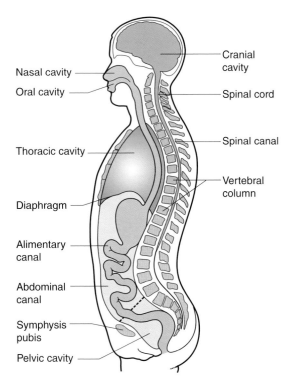

Nasal cavity

Oral cavity

Thoracic cavity

Diaphragm

Alimentary canal

Abdominal canal

Symphysis pubis

Pelvic cavity

Cranial cavity

Spinal cord

Spinal canal

Vertebral column

Fig. 1.5 Diagrammatic section of the cavities of the head and trunk (the unlabelled broken line marks the boundary between the abdomen and the pelvic cavity).

SUMMARY

This chapter has reviewed some of the physical properties of matter and also outlined the anatomical organization of the body. The remainder of this section is concerned with the physiological organization of the body into cells, tissues and organs, with a brief consideration of homeostasis. Further anatomical details will be given in the following sections, which look at the systems of the body.

SELF-TEST QUESTIONS

- How can states of matter be changed?
- What is an acid?
- What does a mole of any substance contain?
- What are the characteristics of living matter?
- Explain the terms 'proximal' and 'distal'.

Chapter 2

The chemical level of organization

CHAPTER CONTENTS

LEARNING OBJECTIVES

After reading this chapter you should understand:

- which three particles comprise an atom
- what the atomic number of an element is
- what the atomic weight of an element is
- how many electrons there are in the first level orbitals
- what the valency of an element is
- what a covalent bond is
- what the primary, secondary, tertiary and quaternary aspects of the structure of a protein are
- what part hydrogen bonds play in protein and DNA structures
- the structure of the double helix of DNA
- the central dogma of genetics

This chapter covers some basic concepts in chemistry, including chemical bonding, and applies this to the range of biological macro-molecules – fats, carbohydrates, proteins and nucleic acids.

THE STRUCTURE OF MATTER

All matter is made up of tiny particles called *molecules* which are so small that they cannot be seen with an ordinary microscope. Molecules are held together only by the attraction of one atom to another, in the same way that a needle is held to a magnet. It is perhaps rather difficult to accept the fact that an iron bar is made of molecules which do not even adhere to each other but are only held together by mutual attraction. It is, however, an important concept that must be remembered. The force of attraction is called cohesion and must be distinguished from adhesion. A molecule is the smallest particle that can exist alone but which still has all the properties of the substance. For example, a molecule of water has all the properties of water which were mentioned earlier.

A molecule can be divided into smaller particles called *atoms* but an atom cannot exist alone. A molecule of an element consists of atoms of that element, e.g. a molecule of oxygen consists of two atoms of oxygen (written O_2). A molecule of a compound consists of atoms of the different elements present in the compound, e.g. a molecule of water consists of two atoms of hydrogen and one atom of oxygen (written H_2O). A molecule of carbon dioxide consists of one atom of carbon and two atoms of oxygen (CO_2) and a molecule of glucose, which is a simple form of sugar, consists of six atoms of carbon, twelve atoms of hydrogen and six atoms of oxygen ($C_6H_{12}O_6$). These are examples of very simple chemical compounds. In the more complicated compounds that form the human body, many elements may be present, and the number of atoms of any one element may run into hundreds. Although all molecules are very small they vary considerably in size and complexity.

Atomic number and atomic weight

The word 'atom' means indivisible and the name was given when scientists thought that the atom could not be divided. Now, however, more is known about the structure of an atom and most people have heard about 'splitting the atom' and its connection with atomic energy. An atom consists of even smaller particles of three different types: protons, which carry a positive electrical charge; electrons, which carry a negative electrical charge; and neutrons, which are neutral. The protons and neutrons are massed together to form the nucleus of the atom, while the electrons are arranged around the nucleus in one or more energy levels. Different elements have different numbers of protons and electrons forming their atoms and the number of protons is equal to the number of electrons. This number is called the *atomic number*. Hydrogen has one proton in its nucleus and one electron circling round it and its atomic number is one. It has no neutron. Carbon has six protons and six neutrons in its nucleus and six electrons, arranged round it, so its atomic number will be six. Sodium has 11 protons and 12 neutrons in the nucleus, and 11 electrons round it, so the atomic number is 11 (Fig. 2.1). These numbers do not have to be memorized because they can be obtained from tables. The presence of neutrons in the nucleus does not affect the atomic number.

The weight of a neutron is roughly equal to the weight of a proton and the number of neutrons added to the number of protons gives the *atomic weight*. The weight of an electron is so small that it is not significant. Hydrogen has an atomic weight of one as it has one proton and one electron but no neutron. Carbon has an atomic weight of 12 as it has six protons and six neutrons in its nucleus. Sodium has an atomic weight of 23 as it has 11 protons and 12 neutrons in its nucleus.

ORGANIC CHEMISTRY

There are rules governing the distribution of electrons between energy levels and these rules help us to understand *chemical bonding*. The maximum number of electrons in the first energy level is two and the maximum number in the second energy level is eight. The electrons are usually represented as circling the nucleus in two orbits. The electrons in the second energy levels, however, are arranged in four *orbitals* which project from the nucleus in such a way as to form the four corners of a tetrahedron (Fig. 2.2).

The rules governing the distribution of electrons between orbitals is that each orbital must have at least one electron and can only have a maximum of two. If these simple rules are applied to carbon then there is a distribution of one *unpaired electron* in each second energy level orbital. Oxygen, with six electrons in the second energy level, has two paired electrons and two unpaired electrons. The number of unpaired electrons is important as this is equal to the

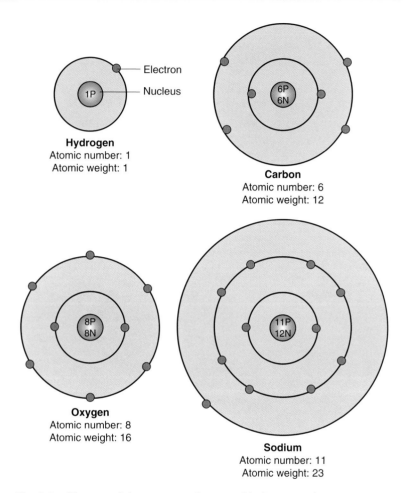

Fig. 2.1 Diagram of the structure of atoms of hydrogen, carbon, oxygen and sodium. P, proton; N, neutron.

valency of the element which dictates how many bonds the element, in its atomic form, can make with other elements. Hydrogen, due to its single unpaired electron in the first energy level, has a valency of one.

Covalent bonding

Elements with unpaired electrons can share these with other similar elements and thereby form covalent bonds. One atom of carbon can

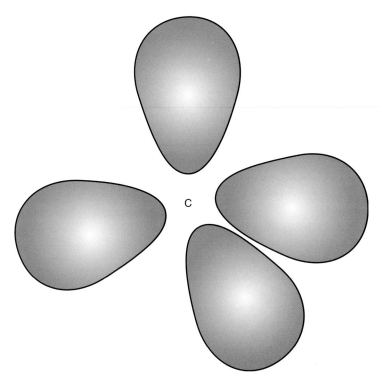

Fig. 2.2 Diagram of the four second-energy-level orbitals of a carbon atom.

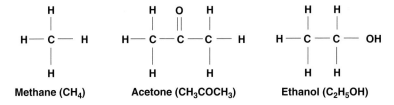

Fig. 2.3 Some simple organic compounds that illustrate the principles of covalent bonding.

form covalent bonds with four atoms of hydrogen to form the simplest organic *hydrocarbon* compound, methane (CH_4). Oxygen can bond covalently with carbon and hydrogen to form acetone (CH_3COCH_3). Another compound containing carbon, hydrogen and oxygen is ethanol (C_2H_5OH) (Fig. 2.3). Covalent bonds are relatively strong bonds that require energy, for example, in the form of heat, to break them. Once they are broken, they release energy and this is why we burn fossil fuels such as coal and oil. It also explains the energy content of foods such as fats and carbohydrates. Nitrogen, which has a valency of three, also forms organic compounds and is incorporated in the amino and nucleic acids which will be described below.

Ions and electrolytes

Atoms do not remain static as electrons may be split off from one atom and gained by another; atoms are constantly gaining and losing electrons. When a substance is in solution in a fluid and an electron splits from an atom of the substance, an ion is formed. Substances that dissociate in this way are known as *electrolytes* because they carry small electric charges. Electrolytes are present in body fluids and are essential to life.

Sodium and chlorine are examples of elements that form *electrolytes*. Sodium readily loses the unpaired electron in its outer energy level giving it a positive charge, and chlorine can accept an electron into its outer energy level, giving it a negative charge. When sodium and chlorine form a compound, the common salt sodium chloride (NaCl), the bond that exists between them is called an ionic bond. Such bonds are easier to break than covalent bonds and this is demonstrated by the fact that salt can be dissolved by adding it to water.

Sometimes when weak chemical bonds break they leave one molecule with an unpaired electron and this results in a free radical. Free radicals react very quickly with other molecules and this results in the formation of another free radical and so on which can lead to damage and even the death of cells. The body can defend itself against free radicals using chemicals called antioxidants.

Hydrogen bonds

The best example of hydrogen bonding is in water (Fig. 2.4) where there is an unequal distribution of the electrons in the covalent bond between the hydrogen and oxygen atoms. The electrons are more attracted to the oxygen, giving it a slight negative charge, and the hydrogen a slight positive charge. Such bonding means that water is a liquid and not a vapour.

There are other examples of hydrogen bonds. These are very weak bonds but they occur in such abundance in some organic compounds,

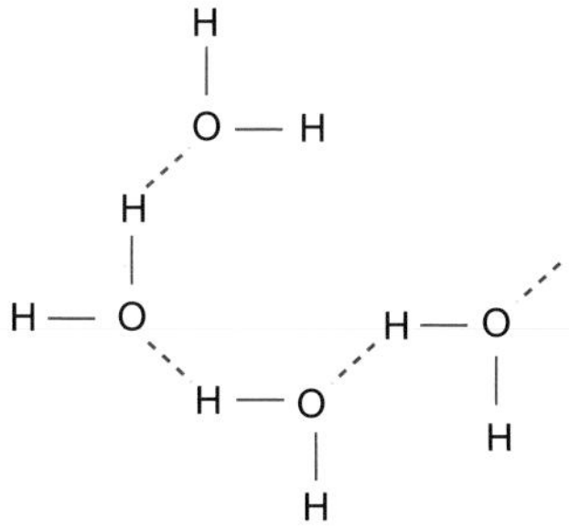

Fig. 2.4 Diagrammatic representation of hydrogen bonding in water. The broken lines denote the weak interactions between the oxygen and hydrogen atoms.

particularly in the biological macromolecules such as proteins and nucleic acids, that they contribute greatly towards the stability of these compounds.

BIOLOGICAL MACROMOLECULES

These include the triglycerides (fats), polysaccharides, proteins and nucleic acids, which have important nutritive, structural and genetic functions. All these molecules are *polymers*, which means that they are built from covalently bound repeating units. Fats are essentially long-chain hydrocarbon molecules attached to glycerol. *Polysaccharides* are built from common carbohydrates (sugars) such as glucose, proteins are built from amino acids, and nucleic acids are built from purine and pyrimidine bases.

Each of these polymers has its own type of covalent bond. The hydrocarbon chains of fats are attached to glycerol by *ester linkages*. The polysaccharides have *glycosidic linkages*, the proteins have *peptide bonds* and the nucleic acids have *phosphodiester bridges*. One of the commonest polysaccharides is starch. All of the enzymes are proteins and DNA (deoxyribonucleic acid), the material from which our genes are made, is a nucleic acid.

Primary, secondary, tertiary and quaternary structure

The biological macromolecules display various levels of structure and, using protein as an example, these can be explained as follows.

Fig. 2.5 The tertiary (folded) structure of a globular protein.

The sequence of amino acids, of which there are 20 in organic compounds, is described as the primary structure.

Specific regions of this primary structure, which are regular in structure due to folding into helices and sheets, are described as the secondary structure, and any subsequent folding is described as the tertiary structure (Fig. 2.5). Where two or more molecules are held together, this is described as the quaternary structure (Fig. 2.6). The primary structure is held together by covalent bonds, and a major factor in the higher order of structure is hydrogen bonding between regions of primary structure, although other forces, including covalent bonding, can maintain these structures.

THE CENTRAL DOGMA OF GENETICS

DNA consists of two strands held together in a double helix by hydrogen bonds (Fig. 2.6). The primary structure of the bases in DNA is the basis of the *genetic code*, which gives rise to all of our characteristics. The code consists of triplets (known as *codons*), each of which leads to the incorporation of an amino acid into a protein molecule.

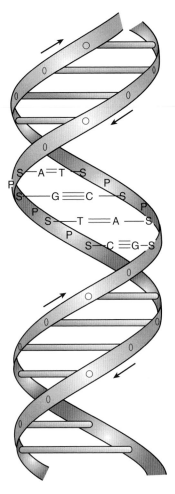

Fig. 2.6   The quaternary structure of DNA is maintained by hydrogen bonds between the two strands.

DNA can *replicate* and conserve the genetic code during cell division (see Chapter 3) and can also be *transcribed* into RNA (ribonucleic acid) (Fig. 2.7). RNA also contains the genetic code and is used as a template for the building of unique protein molecules. This process is, obviously, under very close regulation in living cells in order to ensure that the right protein molecules are made as required. Disturbance of this process and the unregulated production of particular cells is called 'cancer'.

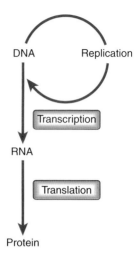

Fig. 2.7 The central dogma of genetics. Replication and transcription take place in the nucleus. Translation takes place in the cytoplasm.

SUMMARY

The main points to be appreciated from this chapter are that biological macromolecules are covalently bonded polymers. The more complex polymers – proteins and nucleic acids – also show weak interactions such as hydrogen bonding between chains.

SELF-TEST QUESTIONS

- What charges do electrons, protons and neutrons carry?
- What is the valency of carbon, hydrogen and oxygen?
- Compare covalent bonds and hydrogen bonds.
- What part do hydrogen bonds play in DNA and protein?
- What are replication, transcription and translation?

Chapter **3**

The cellular level of organization

LEARNING OBJECTIVES

After reading this chapter you should understand:

- the structure of a typical cell
- the structure and function of the plasma membrane
- transport across the plasma membrane
- the essential features of mitosis and meiosis
- different types of cell

The cell is the unit of living matter, the fundamental building block of life, and all cells have certain features in common. This chapter will consider those features and highlight the structure and function of the limiting membrane around the cell and the processes whereby cell division takes place.

STRUCTURE OF A CELL

All cells are made of a substance called *protoplasm*, which is jelly like, opaque and colourless, and consists mainly of water with other substances in solution (Fig. 3.1).

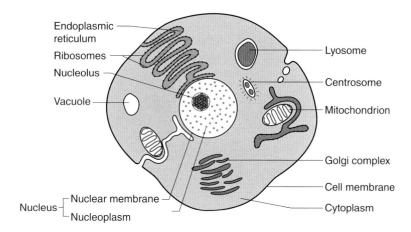

Endoplasmic reticulum
Ribosomes
Nucleolus
Vacuole
Lyosome
Centrosome
Mitochondrion
Golgi complex
Cell membrane
Cytoplasm
Nucleus — Nuclear membrane
Nucleoplasm

Fig. 3.1 Diagram of a cell.

The word *cytoplasm* is commonly used to describe the protoplasm which forms the bulk of the cell, the prefix 'cyto-' being derived from the Greek word for cell. The cytoplasm contains molecules of ribonucleic acid (RNA), which acts as a messenger carrying information out from the nucleus to the cytoplasm.

The plasma membrane plays a very important role in holding the contents of the cell together, and yet is flexible enough to allow cells to conform to many shapes. The plasma membrane also regulates the passage of materials such as nutrients, electrolytes and waste products in and out of the cell. In this way the plasma membrane is described as being *selectively permeable* or semipermeable.

The *plasma membrane* (Fig. 3.2) is composed of protein and a substance called *phospholipid*. The structure of phospholipid molecules, which makes them *hydrophobic* (repelled by water) at one end and *hydrophilic* (attracted to water) at the other end, leads to the unique structure of the plasma membrane, a double layer known as a *phospholipid bilayer*. The protein molecules are embedded in this bilayer and it is the protein molecules that confer the selective permeability on the plasma membrane. The protein molecules can form channels or carriers for the transport of certain substances across the plasma membrane.

Other functions of the protein in the plasma membrane are to act as sites of recognition between cells and to act as receptors whereby substances such as hormones (see Chapter 8) can attach to cells and influence their ability.

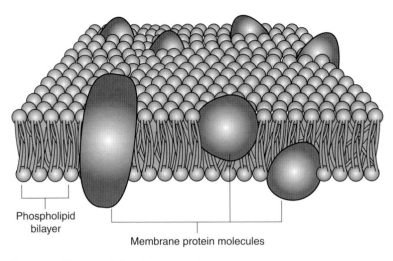

Fig. 3.2 Diagram of the plasma membrane.

TRANSPORT PROCESSES ACROSS PLASMA MEMBRANES

Movement of substances across cell membranes takes place by several processes:

- Osmosis
- Passive diffusion
- Facilitated diffusion
- Active transport.

The processes of osmosis and diffusion, whereby water and substances in solution, respectively, move across semipermeable membranes, have already been described in Chapter 1, and these work across the plasma membrane.

The molecules of some substances are unable to diffuse across the cell membrane. The reasons for this are that the molecules may be too large or may be repelled by the hydrophobic lipid bilayer. One of the processes by which such substances, which the cell requires, can cross the plasma membrane is facilitated diffusion. Facilitated diffusion can only take place where the concentration of the substance outside the cell is greater than the concentration inside the cell. However, there are specific sites in the plasma membrane (protein carriers) at which the substances can cross. Some cells accumulate nutrients such as carbohydrates in this way.

A further process whereby substances can be accumulated or expelled from cells is called active transport. This process requires specific carriers in the membrane but differs from facilitated transport in that it can accumulate substances regardless of their concentrations on either side of the plasma membrane, and it achieves this by using energy. A good example is the way in which cells accumulate potassium (K^+) and expel sodium (Na^+) ions and thereby maintain a high concentration of potassium inside the cell relative to the outside. The cells of the liver accumulate glucose in the same way.

The nucleus is a dense body within the cell. It contains nucleoplasm and is surrounded by a nuclear membrane. The characteristic compounds of the nucleus are deoxyribonucleic acids (DNA), which contain the genetically inherited information required for the maintenance of the cell. The nucleoplasm stores the information necessary for the cell to grow and to divide into two daughter cells. This information is stored in the *genes*, which are folded up to form chromosomes.

Chromosomes are normally only visible under the microscope when the cell is about to divide.

The mitochondria are the power stations of the cell. They are responsible for converting the food ingested by the cell into energy.

Vacuoles are clear spaces within the cytoplasm, which contain waste materials, or secretions, formed by the cytoplasm.

The centrosome is a small rod-shaped body near the nucleus and is important in cell division. It is surrounded by a radiating thread-like structure and contains two centrioles.

CELL REPRODUCTION

Mitosis

In the human body, cell division takes place by a process called mitosis (Fig. 3.3). There are seven stages:

1. The centrosome divides into two and each moves away from the other, though they are still attached by thread-like structures. This stage is called the *prophase*.
2. The nuclear material forms dark thread-like structures, the chromosomes. There are 46 chromosomes in human cells, though the number differs in other species.
3. The nuclear membrane disappears and the chromosomes arrange themselves around the centre of the cell. They appear to be attached to the thread-like structure of the centrosomes,

which are by now at either end of the cell. These two changes are known as the *metaphase*.
4. The chromosomes divide longitudinally into two equal parts.
5. The two groups of chromosomes move away to either end of the cell and arrange themselves around the centrosomes. The

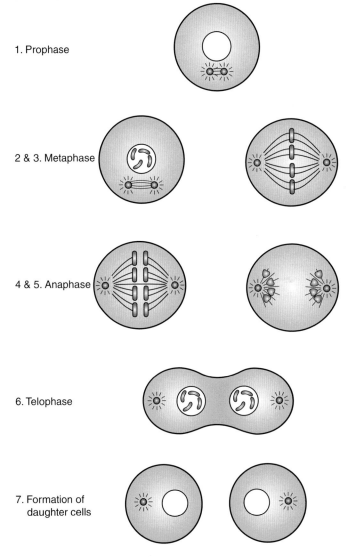

Fig. 3.3 Diagram illustrating mitosis.

thread-like structures joining the centrosomes now divide. These two changes are called the *anaphase*.

6. The cell body becomes narrower round the centre. The thread-like structures disappear and two nuclear membranes reappear. This is called the *telophase*.

7. The cell divides and the chromosomes disappear into the nucleus. This phase, between divisions, is called the *interphase*. The daughter cells will in turn grow and reproduce by mitosis.

The cells produced by mitosis are called *somatic cells*.

Meiosis

Reproduction in higher animals, including man, depends on the fusion of a spermatozoon from the male parent and an ovum from the female. These reproductive cells are also called gametes. Each gamete must receive half as many chromosomes as somatic cells so that when they unite in fertilization the full complement of chromosomes is obtained. As the sex cells mature, two processes of cell division occur; the first is mitosis, in which each daughter cell receives a complete set of chromosomes. Following this a two-stage cell division occurs which is peculiar to reproductive tissue and is called meiosis (Fig. 3.4). The first division is similar to mitosis and gives rise to two

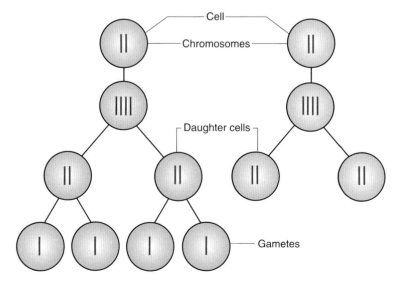

Fig. 3.4 Diagram comparing meiosis and mitosis.

daughter cells, each containing the original number of chromosomes; the second division follows the first very quickly and results in four gametes each containing half the number of chromosomes. In humans normal body cells have 46 chromosomes (in 23 pairs) while each gamete has only 23 single chromosomes. Following fusion of the gametes the resulting cell, a *zygote*, has 46 chromosomes (in 23 pairs). Cell division of a zygote is by mitosis and the resulting multi-celled organism is called an embryo.

The chromosomes consist of a chain of genetic material (DNA), and it is the genes which pass on the characteristics of the parent cell, so that daughter cells are always similar to parent cells. In this way, some characteristics of children are derived from parents, including such characteristics as hair colour, height, degree of intelligence, etc. In any pair of genes, one will exert a stronger influence than the other. The stronger gene is termed *dominant* and the weaker one *recessive*. Hereditary characteristics depend upon the dominance of the genes.

SEX DETERMINATION

One pair of chromosomes from the father and one pair from the mother are the sex chromosomes which will determine the sex of the child. In the female the sex chromosomes are the same and are called XX. In the male they are different and are called XY. One chromosome from each pair will determine the sex of the child. If the child has an X chromosome from the mother and an X chromosome from the father, it will be a girl (XX). If the child has an X chromosome from the mother and a Y chromosome from the father, it will be a boy (XY) (Fig. 3.5).

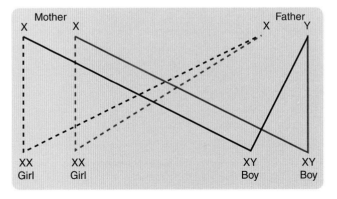

Fig. 3.5   Diagram to show sex determination.

Genetics

Mitosis and meiosis ensure that the genetic information contained in DNA is passed on during the process of cell division and between generations. This is how genetic traits are both inherited between generations and also appear in each cell of the body and are then expressed appropriately, for example, as eye colour or hair colour.

However, this process can go wrong. DNA can become damaged, for example, by radiation or by free radicals (see Chapter 1). The process of replication does not always work perfectly and when damage occurs or replication goes wrong mutations occur. Such mutations can arise during meiosis and can lead to genetic disorders. The nature of the disorder will depend upon where in the genome the mutation has taken place but Down's syndrome, Huntington's chorea and phenylketonuria are classic examples of human hereditary disorders. Nurses are increasingly required to understand genetics and hereditary disorders and some nurses specialize in such work as genetic counsellors working with parents who may be at risk of having a baby with a genetic disorder.

EUKARYOTES

The cells of which the human body and all higher organisms, such as animals, are comprised are called eukaryotic cells. The cells of bacteria are different in some respects and these are called prokaryotic cells; these are characterized by having a nucleus and a plasma membrane.

MULTICELLULAR ORGANISMS

Multicellular organisms consist of many cells. Each cell is living and requires food, oxygen, water, a suitable temperature and the correct pH, but it is dependent on other cells to supply it with the necessities of life in return for the specialized work it carries out. For example, the cells of the lungs take in oxygen, and the cells of the digestive tract absorb food. Each cell develops in a specialized manner so that it can carry out its task satisfactorily. This specialized development is known as the *differentiation* of cells. Groups of specialized cells, developed for a particular purpose, form the tissues of the body, e.g. muscle tissue for movement and bone for support.

The multicellular organism in which we are interested is the human body, a complex arrangement of millions of cells but still having the basic characteristics of all living matter.

SELF-TEST QUESTIONS

- Name five structures found in cells.
- Compare passive, facilitated and active transport processes across the plasma membrane.
- What are the functions of mitosis and meiosis?
- What type of cell is a bacterium?
- What is differentiation?

Chapter 4

Tissues, organs, systems and homeostasis

LEARNING OBJECTIVES

After reading this chapter you should understand:

- definitions of tissues, organs and systems
- classification of tissues
- functions of tissues
- different types of cartilage
- different types of muscle
- how organs may be classified
- the systems of the body
- homeostasis

There is a logic to the way in which the body is organized anatomically and physiologically, and this can best be studied by looking at organizational levels in the body above that of the cellular level. This chapter begins with the next level of organization, that of tissues, where more than one cell comes together to perform a similar function, up to the level of systems. Finally, a major feature of physiology, homeostasis, will be briefly considered.

TISSUES

The body consists of countless cells developed to form various different types of tissue. Tissue originates from a single typical cell, the egg cell or ovum, which is composed of protoplasm and contains a nucleus. After fertilization this cell multiplies and, by the process of differentiation, develops into all the various tissues required to form the different organs and parts of the body.

In the very early stages the ball of cells is divided into three layers. The outer layer is called the *ectoderm*, and from it the outer part of the skin develops with its nails, hair follicles and sweat glands, and other epithelial tissues including the mucous membrane lining the nose and the mouth, plus the enamel covering the teeth. The nervous system also originates in the ectoderm. The middle layer is called the *mesoderm* and from it develop muscle, bone, fat and some of the internal organs, including parts of the cardiovascular system. The inner layer is called the *endoderm* and from it the lining of most of the alimentary and respiratory tracts develop.

A *tissue* consists of cells and the products of cells specially developed for the carrying out of a special function. In the body there are four main types of tissue (Table 4.1):

- Epithelial tissue or epithelium
- Connective tissue
- Muscular tissue
- Nervous tissue.

Epithelial tissue

Epithelial tissue provides covering and lining membranes for the free surfaces inside and outside the body and is the tissue from which the glands of the body are developed. Epithelia protect underlying tissue from wear and tear but must be continually renewed as needed. Some cells are specially developed to be able to absorb

Table 4.1   Classification of tissues

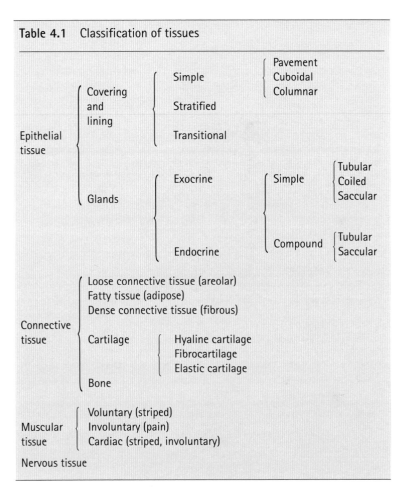

Epithelial tissue	Covering and lining	Simple	Pavement Cuboidal Columnar	
		Stratified		
		Transitional		
	Glands	Exocrine	Simple	Tubular Coiled Saccular
			Compound	Tubular Saccular
		Endocrine		
Connective tissue	Loose connective tissue (areolar) Fatty tissue (adipose) Dense connective tissue (fibrous)			
	Cartilage	Hyaline cartilage Fibrocartilage Elastic cartilage		
	Bone			
Muscular tissue	Voluntary (striped) Involuntary (pain) Cardiac (striped, involuntary)			
Nervous tissue				

substances – these linings are only one cell in thickness and often have a specialized surface called a 'brush border'. Some epithelial tissue, particularly glandular tissue, has the ability to secrete substances manufactured within the tissue. Epithelia do not contain blood vessels, the nearest being in the underlying connective tissue, which may be some distance away. The cells are arranged on a 'basement membrane', which plays a part in binding them together.

Covering and lining epithelia

Covering epithelia may be classified according to the arrangement and shape of their cells.

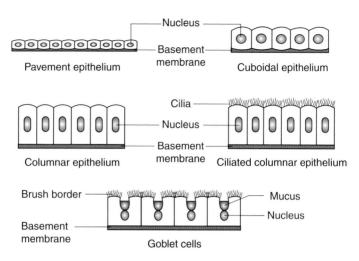

Fig. 4.1   Diagram showing types of simple epithelium.

Simple epithelium is composed of a single layer of cells attached to a basement membrane; it is very delicate and is found where there is little wear and tear (Fig. 4.1).

1. *Simple pavement epithelium* is composed of flat cells, which form a smooth lining. These may be found lining the blood vessels and forming the peritoneum.

2. *Simple cuboidal epithelium* has cells shaped like cubes and is found covering the ovary.

3. *Simple columnar epithelium* is composed of taller cells, packed onto a basement membrane. It is found where wear and tear is a little greater, for example, lining the stomach and intestines. According to the functions performed, columnar epithelium may be modified.

Ciliated columnar epithelium   has microscopic hair-like processes projecting from the free surface of the cell. The cilia move together with a wave motion, and mucus and other particles are moved along. This type of tissue is found in the respiratory tract.

Goblet cells   are cells that secrete mucus, which collects in the cells until the cytoplasm becomes distended.

A brush border   is found on cells specialized for absorption. These have minute finger-like projections that increase the area over which absorption can occur. They are found in the small intestine.

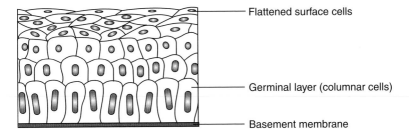

Fig. 4.2   Diagram illustrating stratified epithelium.

Stratified epithelium is made up of many layers of cells (Fig. 4.2). The deepest cells, called the *germinal layer*, lie on the basement membrane and are columnar. As they divide, and this occurs frequently, the parent cells are pushed nearer the surface and become flattened. The cells on the surface are rubbed off and are continually replaced from below. If the surface of the epithelium is dry, as on the skin, the surface cells die because the blood supply is below the basement membrane, and the scaly surface which develops, *keratin*, constitutes a waterproof layer. If the surface is moist, as in the mouth, the cells survive until they are rubbed off, and so keratin is not formed.

Transitional epithelium is like stratified epithelium but the surface cells, instead of being flattened, are rounded and can spread out when the organ expands. Transitional epithelium is found lining organs which must expand and must be waterproof, e.g. the bladder.

Glands

Glands develop from epithelial tissues and have the ability to manufacture substances from materials brought by the blood. These substances are called the *secretions* of the gland. For example, blood contains sodium chloride; from this the gastric glands can make the hydrochloric acid found in the gastric juice, although to obtain hydrochloric acid from sodium chloride in the laboratory is a difficult process. Glands are well supplied with blood vessels, which supply the cells with the materials necessary for making secretions. Glands are of two types: exocrine and endocrine.

Exocrine glands   Exocrine glands (Fig. 4.3) pour their external secretions out through a duct. The secretions of many of these glands contain enzymes, which are chemical substances produced by the

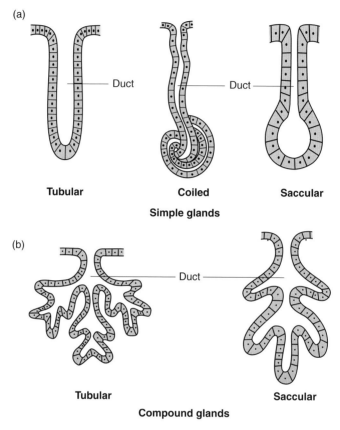

Fig. 4.3   Types of exocrine gland.

cells of the gland. These enzymes produce chemical changes when they come into contact with specific substances but do not themselves enter into the reaction.

Simple glands   Simple glands have one duct leading from a single secretory unit:

1. Simple tubular glands are found in the walls of the small intestine and the stomach.
2. Simple coiled glands pour sweat onto the surface of the skin.
3. Simple saccular glands, known as sebaceous glands, secrete a substance called sebum, which lubricates hair and skin.

Compound glands Compound glands have several secretory units pouring their secretions into a number of small ducts which unite to form a larger duct:

1. Compound tubular glands are found in the duodenum.
2. Compound saccular glands are also called racemose glands; examples are the salivary glands in the mouth.

Endocrine glands Endocrine glands pour their internal secretions directly into the blood stream. These secretions are called *hormones*. Examples of endocrine glands are the pituitary gland in the skull and the thyroid gland in the neck.

Connective tissue

Connective tissue supports and binds together all other tissues. There are many varieties of connective tissue, which differ greatly in appearance, though there is similarity in their connective function and in the fact that all originate from primitive cells called *mesenchyme cells* which themselves originate in the mesoderm. Connective tissue consists of cells, intercellular substance called matrix, and fibres. Matrix and fibres are non-living products of the cells, which form the supporting material of the body. The fibres are of two main types: collagenous and elastic.

Collagenous fibres originate from cells called fibroblasts, which secrete a substance that becomes *collagen*. These coarse fibres occur in wavy bundles and will stretch only a little without tearing.

Elastic fibres are fine branching fibres that are highly elastic.

In the body a connection sometimes needs to be firm and unyielding and sometimes a degree of elasticity is required. For example, the fibrous layers surrounding organs must be somewhat elastic to allow for swelling when the part is engorged with blood, but the fibrous tendons which join muscle to bone must be inelastic, since if they were elastic the tendon would stretch when the muscle contracted and the bone would not move. There are five main varieties of connective tissue.

Loose connective tissue

This type of tissue, also called *areolar tissue*, consists of a loose network of both collagenous and elastic fibres with small scattered groups of fat cells and some fibroblasts (Fig. 4.4). Some blood vessels and nerves are found in the tissue but these are not very numerous. Areolar tissue forms a transparent skin, as thin as tissue paper, but very tough, and it is found between and around the organs of the body.

Fatty tissue

This tissue, also known as *adipose tissue*, is similar to areolar tissue except that the spaces of the network are filled by fat cells (Fig. 4.5). Fat cells contain a globule of fat, which pushes the cytoplasm and the nucleus to the edge of the cell. Adipose tissue is useful because it forms a food reserve on which the body can draw in time of need, for example, in severe exercise or starvation; adipose tissue helps retain body heat because it is a poor conductor of heat and it protects delicate organs such as the eye and the kidney.

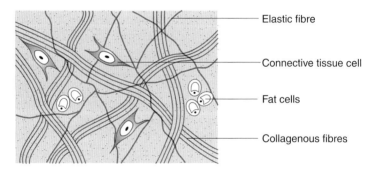

Elastic fibre

Connective tissue cell

Fat cells

Collagenous fibres

Fig. 4.4 Loose connective tissue.

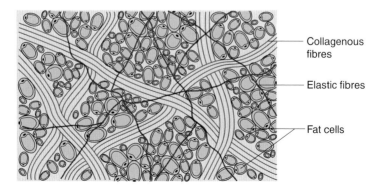

Collagenous fibres

Elastic fibres

Fat cells

Fig. 4.5 Fatty or adipose tissue.

Dense connective tissue

Also called *fibrous tissue*, dense connective tissue consists chiefly of bundles of collagenous fibres between which the fibroblasts lie (Fig. 4.6); it is very strong compared with loose connective tissue. The fibres may be arranged regularly, in parallel bundles (as in tendons or ligaments) or irregularly, with the fibres running in different directions (as in the sheath enclosing muscles), which is called fascia.

Cartilage

Cartilage consists of cells, chondrocytes, separated by fibres. It has no blood vessels so the cells obtain their nourishment by diffusion of fluid through the intercellular substance. Cartilage is very tough but it is also pliant. There are three types (Fig. 4.7):

1. *Hyaline cartilage* consists of chondrocytes embedded in an apparently structureless matrix, which is glassy in appearance and which has very fine collagen fibres running through it. It is found in the trachea and covering the ends of bones at a joint.

2. *Fibrocartilage* contains more collagen fibres than hyaline cartilage and is therefore stronger. It is found between bones forming slightly movable joints, e.g. between the bodies of the vertebrae.

3. *Elastic cartilage* contains numerous elastic fibres embedded in the matrix and can be found, for example, in the epiglottis and the auricle of the ear.

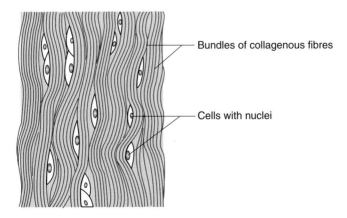

Bundles of collagenous fibres

Cells with nuclei

Fig. 4.6   Diagram of dense connective (fibrous) tissue.

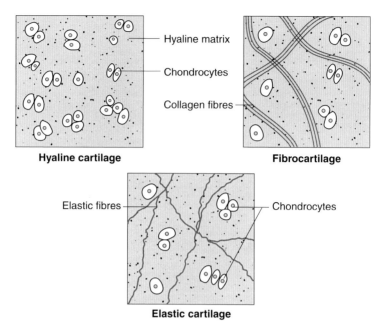

Fig. 4.7 Types of cartilage.

Bone

Bone is a specialized type of cartilage in which the collagen has been impregnated with mineral salts, chiefly calcium. The collagen fibres make the bone tough and the mineral salts make it rigid so that it gives proper support to the soft tissues. The cells between the fibres are called osteocytes and bone is richly supplied with blood vessels. The structure of bone will be dealt with fully in Chapter 10.

Haemopoietic tissue is concerned with the formation of the blood cells and is derived from the primitive mesenchyme cells. Blood may be regarded as connective tissue with the plasma forming the matrix, which contains the cells. It will be dealt with fully in Chapter 14.

Muscular tissue

Muscular tissue is specialized for contraction and is therefore able to produce movement. Wherever there is movement in the body there must be muscular tissue to produce it. Muscle cells are long and thin so that the shortening that occurs during contraction may be as effective as possible. Muscle cells are often called 'muscle fibres'

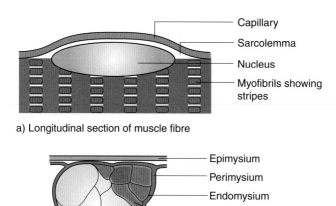

a) Longitudinal section of muscle fibre

b) Cross-section of part of muscle

Fig. 4.8   Sections of skeletal muscle.

because of their shape. The elastic fibres of connective tissue, if they have been stretched, can recoil and return to their original length, but muscle fibres can shorten themselves without this preliminary stretching. There are three types of muscular tissue: skeletal; smooth; and cardiac.

Skeletal muscle

Skeletal muscle forms the flesh of the limbs and trunk, by which the skeleton is moved. It consists of long cells, often known as fibres, that vary in length from a few millimetres in short muscles to 30 cm or more in long muscles; each cell contains numerous thread-like fibres called *myofibrils*, which are only from 0.01 to 0.1 mm in width. These myofibrils are regularly striped crossways in alternate light and dark bands throughout their length and are so arranged that the dark and light parts lie next to one another (Fig. 4.8). Each fibril is enveloped in a sheath of connective tissue called the *sarcolemma*. The fibrils are bound together by connective tissue into bundles, each enclosed in a sheath called the *endomysium*; these bundles are in turn bound together and enclosed in a sheath called the *perimysium*, to form ultimately the individual muscles, which also have a sheath of fibrous tissue called the *epimysium*. The nucleus is at the edge of the cell. This striped muscle is under voluntary control hence it is also known as 'voluntary muscle'. It contracts strongly when stimulated by a nerve fibre but tires quickly. For strong contraction much energy is required so voluntary muscle must have a good blood supply to bring oxygen and nutrients to the cells and to carry away waste products. Capillaries run between individual muscle cells to ensure adequate blood supply.

Smooth muscle

Smooth muscle forms the walls of internal organs such as the stomach, bowel, bladder, uterus and blood vessels. It consists of spindle-shaped cells, each of which contains a nucleus. It is also called 'unstriped muscle'. The cells do not show any stripes and have no sheath but are bound together by connective tissue to form the walls of the various organs (Fig. 4.9). They are not under the control of the will and act without any conscious effort or knowledge. They contract automatically but are supplied by autonomic nerves which regulate their contractions. This type of muscle is designed for slow contraction over a long period and it does not tire easily.

Cardiac muscle

Cardiac muscle is both involuntary and irregularly striped. It is only found in the heart wall and is different from any other muscle tissue. It consists of short, cylindrical, branched fibres with centrally placed nuclei. They have no sheath but are bound together by

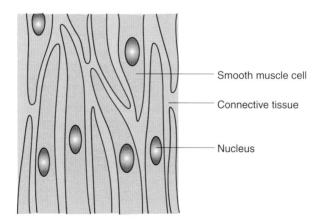

Fig. 4.9   Section of smooth muscle.

connective tissue (Fig. 4.10). The cardiac muscle is not under the control of the will but contracts automatically in a rhythmic manner throughout life, though the rates of these rhythmic contractions are controlled by nerves, which quicken or slow down its action. The fibres branch and join with one another so that impulses can spread from one fibre to another as well as along the length of the muscle.

Nervous tissue

Nervous tissue is specially designed to receive stimuli from either inside or outside the body and, when stimulated, to carry impulses rapidly to other tissues. Nervous tissue consists of nerve cells, called *neurones* (Fig. 4.11), and a special supporting network called *neuroglia*. A nerve cell consists of a large cell body to which several short processes, *dendrites*, bring impulses from other cells and tissues. From the cell body there is one long process called the axon, which carries impulses away from the cell body.

MEMBRANES

The cavities and hollow organs of the body are lined with membranes. These membranes consist of epithelium and secrete lubricating fluids to moisten their smooth, glistening surfaces and prevent friction. Three different types of membrane are found in the body.

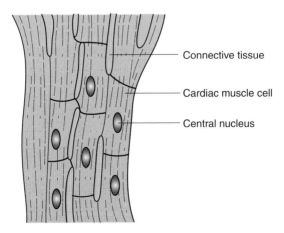

— Connective tissue

— Cardiac muscle cell

— Central nucleus

Fig. 4.10 Section of cardiac muscle.

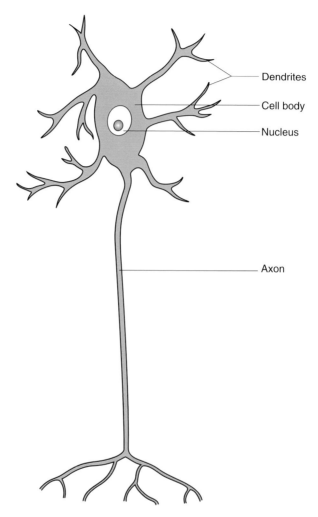

Fig. 4.11 Diagram of a neurone.

Synovial membrane

Synovial membrane secretes a very thick fluid, which is like white of egg in consistency, hence the name: the prefix 'syn-' meaning 'like' (from the Greek) and 'ovum' being Latin for an egg. It is chiefly found lining joint cavities to lubricate the movement of the bones on one another. It is a fibrous membrane covered with epithelium. It is also found over bony prominences and between ligaments and bones or between tendons and bones. In these sites it forms small sacs called *bursae*, which act as water cushions, facilitating the movement of one part on another. For example, there are bursae around the shoulder, knee and elbow joints. Synovial membrane is also found forming sheaths for tendons, which run a long distance to their insertions. For example, the tendons and muscles in the forearms and legs run across the hand and foot respectively and move the fingers and toes.

Mucous membrane

Mucous membrane secretes a rather thinner fluid called mucus. This membrane is found lining the food canal from mouth to rectum and the air passages from the nose downwards. Thus the cavities that it lines are connected with the external skin. Mucus-secreting tubular glands are also present where much secretion is produced. These are single or branching tubes lined with secreting cells.

Serous membrane

Serous membrane consists of flattened cells through which oozes a small quantity of thin watery fluid called *serous fluid*; this is similar to the fluid that oozes from a blood clot. Serous membrane is found lining the internal cavities, e.g. the thorax and abdomen, and covering the organs they contain, providing smooth, glistening, moist surfaces that slide easily over one another when one part moves on, or within, another.

ORGANS

Where two or more tissues exist together to serve a common purpose, such as digestion or movement, the composite structure is described as an *organ*. Examples of organs that relate to digestion and movement are the stomach and the leg, respectively. The stomach is composed of several tissue types including smooth muscle, endothelial, glandular and nervous tissues. The leg is composed of bone, muscle, skin and nervous tissue. All organs have a blood supply.

Organs are described as being either tubular or compact. Tubular organs have certain features in common. They all contain a space called a *lumen* and the structure of the tubular structure can be broadly divided into three layers: an outer epithelial layer, a middle layer containing muscle, and an inner endothelial layer. While no two tubular organs are identical in their structure, these features can be seen in organs as diverse as the heart, which pumps blood round the circulatory system, and the small intestine, which transports food from the stomach to the large intestine. Each organ has unique layers, for example, in the heart the outer layer is called the pericardium, the middle layer is the myocardium and the inner layer is the endocardium. The lumen of the heart forms the chambers called *atria* and *ventricles*.

Compact organs, such as the liver and kidney, have superficial layers called the cortex and deeper layers called the medulla. These organs have no lumen.

SYSTEMS

Where organs of the body work together towards a common purpose, such as circulation, digestion or movement, this is described as a *system*. These systems will be the subject of the remainder of this book and examples of such systems are given below:

- The *skeletal system* provides a framework which gives support and protection to the soft tissues and allows movement at joints.
- The *muscular system* effects movement of the body as a whole.
- The above two systems together are sometimes known as the *musculoskeletal system*.
- The *circulatory system* is the transport system of the body; it carries oxygen and nourishment to the tissues and waste products away from them and is essential if there is to be dependence of one tissue on another.
- The *respiratory system* allows exchange of gases between the body and the environment.
- The *digestive system* is concerned with digestion and absorption of food and the elimination of waste matter.
- The *endocrine system* produces hormones, which control a variety of bodily functions.
- The *urinary system* is the main excretory system of the body.
- The *nervous system* creates awareness of the environment and makes it possible for the body to respond to change with the required precision.
- The *reproductive system* is responsible for the survival of the species by reproduction of the same kind.

HOMEOSTASIS

The concept of *homeostasis* is fundamental to an understanding of physiology. The word 'homeostasis' is derived from a Greek expression meaning 'staying the same'. In physiological terms it means that, despite changes in the external environment of the body, under normal healthy conditions the internal environment of the body remains the same. All the systems of the body work towards this end, and often several systems work together to maintain the constancy of a single aspect of the working of the body.

Extracellular and intracellular environment

One of the best examples of homeostasis is the way in which the body maintains a constant intracellular and extracellular environment with respect to water and electrolyte content. It has already been outlined (Chapter 3) how cells tend to accumulate potassium and expel sodium. Even when the body is at rest, it is using energy, and most of this energy is being used by the active transport processes of cells in order to undertake this exchange of electrolytes across the plasma membrane. Other systems (discussed in later chapters) such as the *urinary system* (kidneys and bladder) are also working to expel or conserve electrolytes. The urinary system and the *respiratory system* work together to maintain the pH of the blood.

Negative feedback

Homeostasis is enabled by means of negative feedback. The components of a negative feedback system are threefold: a detector; an effector; and a control area (Fig. 4.12). The detector senses a change in the internal environment of the body, the effector institutes a change in order to return the internal environment to homeostasis, and the control area determines what the normal physiological situation should be by decreasing the action of the effector once homeostasis has been achieved. Therefore, the effect of the corrective

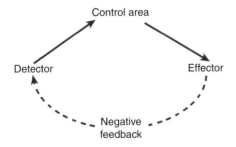

Fig. 4.12   The concept of negative feedback.

change in the system is to stop the change taking place, which is described as negative feedback.

A very good example of homeostasis (described in Chapter 16) is the regulation of blood pressure. In this case the detector is the system of pressure receptors (*baroreceptors*) in the cardiovascular system. The effector is a combination of actions of the heart and blood vessels, and the control area is the *hypothalamus* of the brain. The link between the detector, the effector and the control area is either the nervous system or the endocrine system. If blood pressure, for example, decreases then the heart and blood vessels act to increase the blood pressure in order to return it to normal and once it is normal the actions of the heart and blood vessels are stopped by negative feedback. The internal environment of the body is always changing as we go about our daily lives and, in fact, homeostasis is constantly making adjustments as the internal environment changes.

Positive feedback

Sometimes the body uses positive feedback, which leads to a greater change away from normality once a change has been initiated. One example is blood clotting (Chapter 14) where the stimulus of a damaged blood vessel initiates a cascade of reactions which do not stop until a blood clot has been formed.

Disturbance of homeostasis

Ill health, such as infection or injury, and many medical and surgical procedures result in disturbances of homeostasis. It can be said that nursing begins where homeostasis fails since many nursing procedures and actions are designed to detect such changes and restore balance to the internal environment of the body.

SELF-TEST QUESTIONS

- What are tissues, organs and systems?
- What does glandular tissue do?
- Give two examples of a compact organ.
- Give two examples of a hollow organ.
- How many systems of the body can you name?
- What are the components of a homeostatic system?

SECTION 2

Control and co-ordination

SECTION INTRODUCTION

Everything that the body does, including homeostasis, is controlled by the nervous and endocrine systems, and often these work together to control the internal environment. The nervous system is highly complex and a detailed consideration of the anatomical organization of the brain and spinal cord, collectively known as the central nervous system, and its links to the internal organs and musculature of the body, via the peripheral nervous system, is given.

A particular section of the peripheral nervous system, which is concerned with involuntary actions, the autonomic nervous system, is also considered. Sensory information is relayed to the nervous system by different means, and two principal sets of organs, the eyes and the ears, are covered in specific chapters.

In addition to nervous signals in the body, which involve a combination of chemical and electrical signals between nerves, there is a system that is mediated wholly by means of chemical signals in the blood. The chemicals are called hormones and the organs that secrete these are called endocrine organs. These are, ultimately, controlled by the brain and the system whereby these are regulated and the specific action of some important hormones, such as thyroxine, are described.

Chapter 5

The nervous system

CHAPTER CONTENTS

LEARNING OBJECTIVES

After reading this chapter you should understand:

- the types of nervous tissue
- the difference between myelinated and non-myelinated neurones
- the function of the synapse
- the generation of the action potential
- the components of the central nervous system
- the components of a motor system
- the purpose of the sensory system
- the structure and function of the meninges
- the peripheral nervous system
- the branches of the autonomic nervous system

The nervous system is:

1. The system of communication between the various parts of the body

2. The mechanism by which sensations of all kinds are received from the environment and from the tissues and organs of the body itself

3. Responsible for the interpretation of these sensations through dependence on association with similar sensations received in the past and stored in the memory

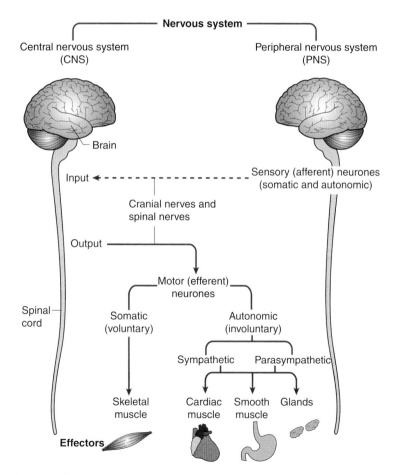

Fig. 5.1   The divisions of the nervous system.

4. The system by which actions are carried out by the sending of impulses to other parts of the nervous system and other organs of the body.

Figure 5.1 shows how the nervous system is organized into the central and peripheral nervous systems and also shows what these divisions of the nervous system are responsible for.

NERVOUS TISSUE

When nervous tissue is examined under the microscope it is seen to be composed of:

- Neurones – nerve cells with associated nerve fibres
- Neuroglia – a special type of connective tissue, found only in the nervous system, which supports the neurones.

The neurone is the basic tissue of the nervous system. It has a comparatively large cell body, though both shape and size vary according to the position of the cell and its function. Each has a clearly defined nucleus and the protoplasm is granular. Nerve cells form the *grey matter* of the brain and spinal cord. The cell has several processes: *dendrites* are short branched processes through which nervous impulses enter the cell; and the *axon* (or axis cylinder), which is a single fibre through which impulses pass out of the cell. Axons vary considerably in length from a few millimetres to many centimetres and are continuous from cell to termination. Nerve processes, or fibres, form the *white matter* of the brain and spinal cord.

A neurone with many processes arising from the cell body is termed a *multipolar* neurone (Fig. 5.2). Other types of neurone have one process arising from the cell body, which then divides into two branches, one – the axon – conveying impulses towards the central nervous system, and one conveying impulses from an organ to the cell; these are *unipolar neurones* (Fig. 5.3). Bipolar neurones have two processes, one at each end of the cell; one is a dendrite carrying impulses to the cell and one is an axon conveying impulses from the cell.

Axons and some dendrites are surrounded by a thin fatty sheath composed of myelin, which lies inside the outer covering of connective tissue, which is called the *neurilemma* (Fig. 5.4). The myelin sheath is compressed at intervals and here the neurilemma dips in

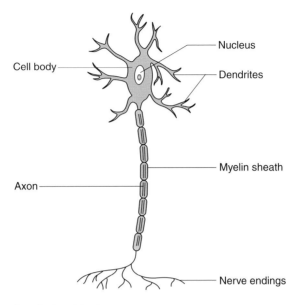

Fig. 5.2   A typical multipolar neurone.

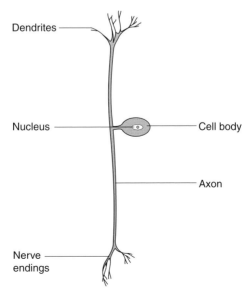

Fig. 5.3   A unipolar neurone.

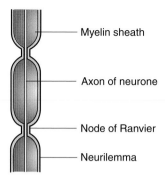

Fig. 5.4   Myelin sheath.

towards the nerve fibre; these constrictions are called the *nodes* of
Ranvier. At these points the nerve fibre has contact with the surrounding tissue fluids, allowing exchange of nutrients and waste
materials.

The myelin sheath is thought to have an insulating effect on the
nerve fibre so that impulses are not transmitted to adjacent nerves
or tissues except through the end of the fibre. The myelin sheath
also protects the fibre from pressure and injury; fibres which have
a myelin sheath are called *myelinated fibres*. Degeneration of the
myelin sheaths around neurones in the spinal cord, optic nerve and
brain is a feature of multiple sclerosis. This condition leads to double
vision, muscular weakness and an unsteady gait.

Non-myelinated fibres are found in the autonomic nervous system
and in certain parts of the brain and spinal cord.

Nerve cells are quickly damaged by lack of oxygen, and by toxins
and poisonous substances. If they die they cannot be replaced,
though it may be possible for the function to be taken over by other
cells to a limited extent.

A synapse is the point of communication between one neurone
and the next. The fibrils forming the axon have tiny expanded ends
called end feet which are close to, but not touching, the dendrites
or cell bodies of other neurones. They allow the passage of the *nerve
impulse* in one direction only.

Also, nerve impulses can only travel in one direction, into a neurone through the cell body or dendrites and out through the axon.
At the synapse there is a short pause to enable a *chemical messenger*
to be released to fill the gap between the two neurones and to allow
the impulse to pass to the next neurone.

THE ACTION POTENTIAL

The distribution of electrolytes across all cell membranes is such that there is a potential difference (voltage) across the membranes. This potential difference is about -70 millivolts (mV). The cells of nerves and muscles are specialized such that they can undergo reversible changes in this potential difference, called an action potential, and this is the basis of their ability to generate and conduct electrical impulses.

When an action potential is generated in a nerve cell by the arrival of an electrical impulse from another nerve cell, positively charged sodium ions (Na^+) enter the cell and reverse the potential difference across the membrane to a slightly positive potential. This process is called *depolarization*. The positive potential difference is rapidly reversed (in less than a millisecond), but not before the process is repeated along the nerve cell, leading to conduction of the electrical impulse. The changes in potential difference are shown in Fig. 5.5.

Action potentials are not graded; in other words, they are not of different sizes in any particular nerve cell. They obey the 'all or none'

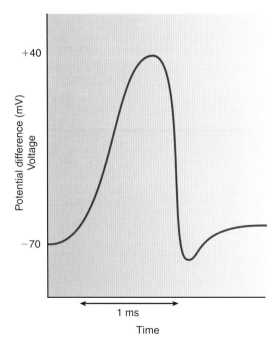

Fig. 5.5   The action potential.

law. Once a certain level of depolarization has taken place through the entry of sodium ions into the cell, known as the *threshold potential*, a full action potential is generated. The size of action potentials differs between nerve cells, and the extent of stimulation in a nerve (which is composed of many nerve cells) depends on the number of nerve cells that have been stimulated. A further feature of nerve cells (and muscle cells) with respect to action potentials is that they display a refractory period of about 0.5 milliseconds during which time another action potential cannot be generated.

The nervous system is composed of a central nervous system, consisting of the brain and spinal cord, and a peripheral nervous system, consisting of the cranial and spinal nerves and the autonomic nervous system.

CENTRAL NERVOUS SYSTEM

The brain

The brain, when fully developed, is a large organ which fills the cranial cavity. Early in its development the brain becomes divided into three parts known as the *forebrain*, *midbrain* and *hindbrain* (Fig. 5.6).

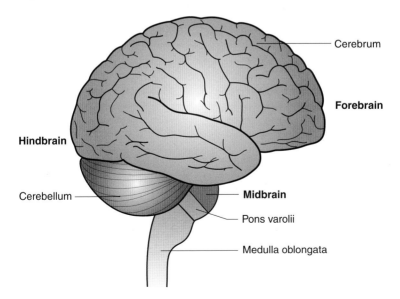

Fig. 5.6   Parts of the brain.

The forebrain is the largest part and comprises the cerebrum; it is divided into right and left hemispheres by a deep *longitudinal fissure* (Fig. 5.7). The separation is complete at the front and the back but in the centre the hemispheres are joined by a broad band of nerve fibres called the *corpus callosum*. The outer layer of the cerebrum is called the *cerebral cortex* and is composed of grey matter (cell bodies) thrown into numerous folds or convolutions called *gyri*, separated by fissures called *sulci*. This enables the surface area of the brain, and therefore the number of cell bodies, to be greatly increased. The general pattern of the gyri and sulci is the same in all humans; three main sulci divide each hemisphere into four lobes, each named after the skull bone under which it lies. The *central sulcus* runs downwards and forwards from the top of the hemisphere to a point just above the lateral sulcus; the *lateral sulcus* runs backwards from the lower part of the front of the brain; and the *parieto-occipital sulcus* runs downwards and forwards for a short way from the upper posterior part of the hemisphere. The lobes of the hemispheres are the *frontal lobe*, lying in front of the central sulcus and above the lateral sulcus; the *parietal* lobe,

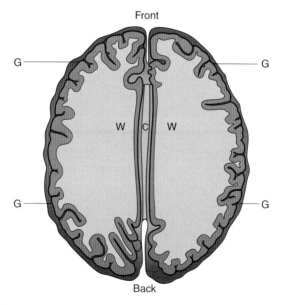

Fig. 5.7 Section through the cerebrum, viewed from above. G, the grey matter of the convoluted surface; W, the white matter forming the central portion; C, the corpus callosum, the bridge of white matter joining the two hemispheres of the cerebrum.

lying between the central sulcus and the parieto-occipital sulcus and above the line of the lateral sulcus; the *occipital lobe*, which forms the back of the hemisphere; and the *temporal lobe* lying below the lateral sulcus and extending back to the occipital lobe (Fig. 5.8).

The area lying immediately in front of the central sulcus is known as the *pre-central gyrus* and is the motor area from which arise many of the motor fibres of the central nervous system. Immediately behind the central sulcus lies the sensory area, called the *post-central gyrus*, in the cells of which several kinds of sensation are interpreted.

A longitudinal section of a hemisphere shows grey matter (cell bodies) on the outside and white matter (nerve fibres) forming the interior (Fig. 5.9). The nerve fibres connect one part of the brain with other parts and with the spinal cord, however within the white matter groups of nerve cells can be seen forming areas of grey matter. These areas of grey matter are called *cerebral nuclei* (Fig. 5.10). The main function of these areas is the co-ordination of movement and posture of the body; disorders affecting these areas cause jerky movements and unsteadiness.

The cavities within the brain are called *ventricles*. There are two lateral ventricles, a central third ventricle and a fourth ventricle between the cerebellum and the pons. All are filled with cerebrospinal fluid.

The midbrain lies between the forebrain and the hindbrain. It is about 2 cm in length and consists of two stalk-like bands of white matter called the *cerebral peduncles*, which convey impulses passing

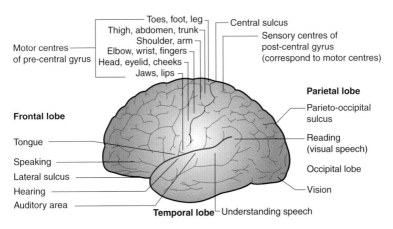

Fig. 5.8 The cerebrum, showing the lobes and the main nerve centres.

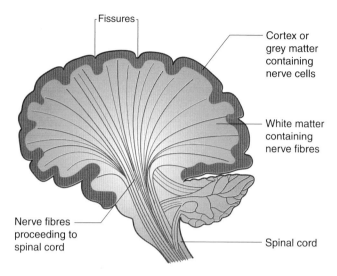

Fig. 5.9 Diagrammatic section of the brain, showing the grey matter on the surface and the white matter in the centre. The association fibres in the cerebrum are not shown.

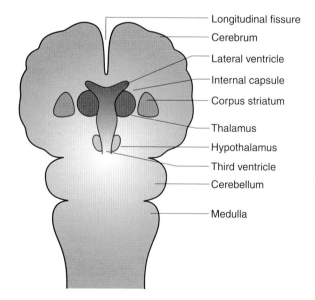

Fig. 5.10 Cerebral nuclei.

to and from the brain and spinal cord, and four small prominences called the *quadrigeminal bodies*, which are concerned with sight and hearing reflexes. The *pineal body* lies between the two upper quadrigeminal bodies.

The hindbrain has three parts:

1. The *pons*, which lies between the midbrain above and the medulla oblongata below, contains fibres which carry impulses upwards and downwards and some which communicate with the cerebellum.

2. *The medulla oblongata* lies between the pons above and the spinal cord below. It contains the cardiac and respiratory centres, which are also known as the vital centres and which control the heart and respiration.

3. The *cerebellum* projects backwards beneath the occipital lobes of the cerebrum. It is connected to the midbrain, the pons and the medulla oblongata by three bands of fibres called the *superior, middle* and *inferior cerebellar peduncles*, respectively. The cerebellum is responsible for the co-ordination of muscular activity, control of muscle tone and maintenance of posture. It is continuously receiving sensory impulses concerning the degree of stretch in muscles, the position of joints and information from the cerebral cortex. It sends information to the thalamus and the cerebral cortex.

The midbrain, the pons and the medulla have many functions in common and together are often known as the brain stem. This area also contains the nuclei from which the cranial nerves originate.

Spinal cord

The spinal cord is continuous with the medulla oblongata above and constitutes the central nervous system below the brain. It commences at the foramen magnum and terminates at the level of the first lumbar vertebra; it is about 45 cm in length. At its lower end, it tapers off into a conical shape called the *conus medullaris*, from the end of which the *filum terminale* descends to the coccyx, surrounded by nerve roots called the *cauda equina* (Fig. 5.11). The cord gives off nerves in pairs throughout its length. It varies somewhat in thickness, swelling out in both the cervical and lumbar regions, where it gives off the large nervous supply to the limbs. The cord is deeply

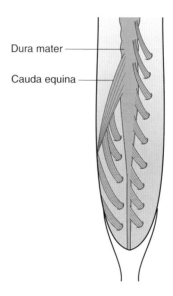

Fig. 5.11   The cauda equina.

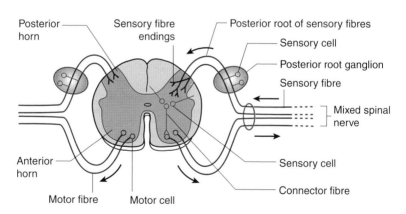

Fig. 5.12   A section of the spinal cord, showing the central grey matter and the neurones. Sensory fibres of heat and pressure cross the cord.

cleft back and front, so that it is almost completely divided into right and left sides like the cerebrum.

The spinal cord, like the medulla, consists of white matter on the surface and grey matter in the centre. The white matter consists of fibres running between the cord and the brain only, not to the body tissues (Fig. 5.12). The cord contains:

- *motor fibres*, which run down from the motor centres of the cerebrum and the cerebellum to the motor cells of the cord
- *sensory fibres*, which run up the cord from the sensory cells of the cord to the sensory centres of the brain.

The grey matter, on cutting across the cord, has an H-shaped pattern, with two portions projecting forwards, one on either side, called the *anterior horns*, and two portions projecting backwards, one on either side, called the *posterior horns*.

The *cranial nerves* constitute 12 pairs of nerves which arise in nuclei in the brain stem. Some are purely sensory, some purely motor and some are mixed nerves carrying both sensory and motor impulses.

The *spinal nerves* constitute 31 pairs of nerves which arise in the spinal cord. Each has a motor and a sensory component in the anterior and posterior parts of the cord respectively, and the two fibres travel together after they leave the cord.

The *autonomic nervous system* is concerned with the control of internal organs; the function of these organs is not under the control of the will, though they are affected by the emotions.

The cranial and spinal nerves and the autonomic nervous system, which as mentioned above form the peripheral nervous system, are discussed in greater detail later in this chapter.

The motor system

The motor system is concerned with the movement of various parts of the body. As already mentioned (see p. 67), the *motor area* is situated in front of the central sulcus in an area called the *pre-central gyrus*. Here the body is represented upside-down. At the bottom of the gyrus is a large area for the head and eye; above that is a large area for the hand and arm, then a small area for the trunk and a large area for the leg extending over the top of the cerebral hemisphere (see Fig. 5.8). There is considerable overlap between these areas. It will be apparent that an area that can undertake a great deal of fine movement such as the hand and arm, will require a larger area of cell bodies than an area such as the trunk which, although greater in extent, does not carry out as much detailed movement. In front of the motor area lies the *pre-motor area* which is concerned with a whole pattern of movement.

Beginning from cells in the motor area the corticospinal fibres pass downwards in a fan shape (see Fig. 5.9) and then pass through the *internal capsule*, which lies between the thalamus and the basal ganglia. All the motor fibres serving one side of the body are gathered together in the internal capsule so injury there will cause paralysis in the affected side (hemiplegia). The fibres then pass through the pons to the medulla oblongata where they form a long narrow projection called a *pyramid*. In the pyramids, most of the motor fibres cross over to the other side, at the *decussation of the pyramids*, so that fibres that arose in the left cerebral hemisphere will now be on the right side of the tract and will supply the right side of the body. The fibres then pass down the spinal cord as the *lateral corticospinal tract*. The fibres that did not cross to the opposite side at the decussation of the pyramids pass down the spinal cord in the *anterior corticospinal tract* and eventually cross to the opposite side in the spinal cord.

The fibres pass to the anterior horn of the spinal cord, where they form a synapse with the cell bodies situated there, and then emerge from the front of the spinal cord as the *anterior root* and join the corresponding posterior root of sensory fibres to form a *mixed spinal nerve* (see Fig. 5.12). As peripheral nerves, these end in branches to various areas, including muscles. The motor fibres divide into branches and each branch ends in a *motor end plate* attached to an individual muscle fibre (Fig. 5.13). Sensory fibres have their cells in the posterior root ganglion on the spinal nerves and have endings of various types within the muscles.

The motor area receives impulses from many other parts of the brain, including the sensory area. From the cortex, impulses are sent to the spinal cord, the motor nuclei in the brain stem, the basal ganglia, the cerebellum and the pons. Through the various nerve tracts,

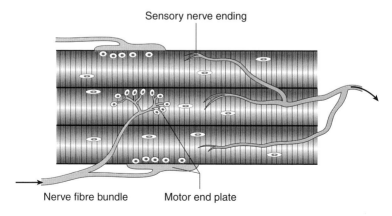

Sensory nerve ending

Nerve fibre bundle   Motor end plate

Fig. 5.13 Muscle, showing motor end plates and sensory nerve endings.

stimuli are passed through the peripheral nerves to the skeletal muscles, which are kept in a state of tension called *muscle tone*. Muscle fatigue is prevented by the use of different sets of motor fibres successively and the degree of tone depends on the number of fibres being used at any one time.

The term upper motor neurone describes the motor fibre within the central nervous system as far as its synapse with an anterior horn cell. A lower motor neurone describes an anterior horn cell and its fibre. Damage to the upper and lower motor neurones leads to lack of control and reflexes, resulting in paralysis. However, after damage to the upper motor neurones the reflexes are eventually restored but are much stronger, resulting in spastic paralysis. In lower motor neurone damage the muscles have no reflexes, which results in flaccid paralysis whereby the muscles atrophy.

The corticospinal tract used to be called the pyramidal tract, so the term *extrapyramidal system* describes all motor systems other than the corticospinal and corticonuclear tracts. The main function of the system is co-ordination of muscle movement in the maintenance of body posture so that movements can be performed accurately while still maintaining the desired posture.

The sensory system

The sensory system is concerned with interpreting the impulses which are constantly stimulating it. Many of these do not reach the level of consciousness because the nervous system deals with them automatically, making adjustments to the blood pressure, the rate of the heart beat, the degree of tone in the muscles and many other conditions.

Sensory impulses are transmitted to the central nervous system from the special sense organs, the skin, and from deeper parts of the body.

Special senses

The impulses of sight are conveyed from the retina of the eye through the *optic nerve* (the second cranial nerve) to the *optic chiasma* where the medial fibres of each optic nerve cross over to the opposite side (Fig. 5.14). Because of this partial crossing the visual area of the left cerebral hemisphere receives impressions from the outer (temporal) side of the retina of the left eye and from the inner (nasal) side of the retina of the right eye. The impulses are then conveyed to

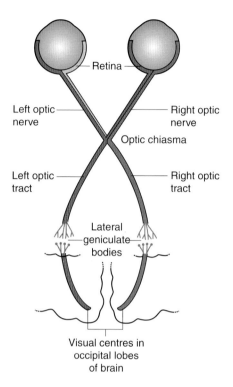

Retina

Left optic nerve

Right optic nerve

Optic chiasma

Left optic tract

Right optic tract

Lateral geniculate bodies

Visual centres in occipital lobes of brain

Fig. 5.14 The visual path from the eyes to the visual centres.

the *visual areas* in the occipital lobes, where they are interpreted. It will be seen that division of the left optic nerve causes blindness of the eye on the same side but that division of the left optic tract causes inability to see the left half of the normal field of vision.

The impulses of hearing are conveyed through the *vestibulo-cochlear nerve* (the eighth cranial nerve) to the pons where there is a synapse. A second set of fibres carries the impulses to the *medial geniculate body* and a third set to the *auditory areas* in the temporal lobes for interpretation.

The impulses of smell are carried to the *olfactory nerve* (the first cranial nerve) and then to the *olfactory bulb* and *olfactory tract*, which are under the frontal lobe (Fig. 5.15). Smells are interpreted in various parts of the temporal lobe. There are few tastes that can be detected without the sense of smell. Taste impulses are carried by the facial (seventh cranial) and glossopharyngeal (ninth cranial) nerves and are interpreted in the temporal lobe with the corresponding smell.

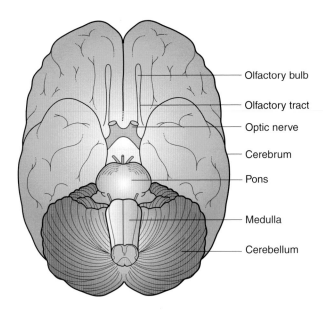

Fig. 5.15   The brain, viewed from below.

Sensation from skin and muscle

Sensory nerve endings are present in the skin and other tissues. Different kinds of sensation – temperature, touch, pressure and degrees of stretch – require different nerve endings to initiate them. In the skin, a sensory nerve fibre may begin as a nerve ending capable of transmitting (1) pain and temperature change, (2) light touch or (3) firm pressure. In addition, muscles have special structures called *muscle spindles* which respond to the degree of tension to which they are exposed. From all these nerve endings, sensory fibres run in the spinal nerves to the posterior nerve roots in the spinal cord. The nerve cell of the first neurone is in the posterior root ganglion. In the head they run in the trigeminal nerve (fifth cranial) and other cranial nerves to the brain.

Nerve fibres carrying light touch and muscle stretch sensations (proprioception) then give off several branches to anterior and posterior horn cells so that each section of the cord becomes a functioning unit. They then pass upwards in the posterior part of the white matter and end in a synapse in the medulla oblongata. A second neurone crosses over to the opposite side and ends at the thalamus. A third neurone carries impulses to the sensory area in the parietal lobe.

Fibres carrying pain and temperature change synapse in the posterior horn. A second neurone crosses over immediately to the opposite side of the cord and ascends to form a synapse in the thalamus. A third neurone passes to the sensory area.

Fibres carrying firm pressure impulses also synapse in the posterior horn. A second neurone crosses over to the opposite side of the cord and ascends, in a different part of the cord, to the thalamus. A third neurone passes to the sensory area.

It can be seen that all sensory fibres eventually cross to the opposite side so that sensation from the left side of the body will be interpreted in the right side of the brain. Also, all sensory neurones form a synapse in the thalamus.

The *thalamus* (see Fig. 5.10) is responsible for sorting out the mass of sensory information being fed into the body and passing it on to the cerebral cortex, when necessary, or to other areas of the brain as appropriate. The *hypothalamus* is concerned with the stability of the internal organs of the body. It controls water balance, regulates appetite, temperature and sleep, and plays a part in the control of emotion.

The sensory area of the cortex lies in the parietal lobe behind the central sulcus. Like the motor area the body is represented upside-down with large areas for the face, head and hands at the lower end and the smaller areas for the arms, trunk and legs, which are comparatively less sensitive.

The meninges

The meninges are the protective membranes covering the central nervous system (Fig. 5.16). There are three layers:

- Dura mater – outer layer
- Arachnoid mater – middle layer
- Pia mater – inner layer.

The dura mater is a tough fibrous membrane with two layers, the outer layer lining the inner surface of the skull and forming the periosteum. At the foramen magnum, this layer continues as the periosteum on the outer surface of the skull. The inner layer of dura projects inwards in certain places to form a double layer, which separates parts of the brain and helps to maintain them in position. The *falx cerebri* is one such fold, between the two cerebral hemispheres, another is the tentorium cerebelli between the cerebrum and the cerebellum. The two layers of the dura are in contact with one another for the most part but are separate when they enclose a venous sinus.

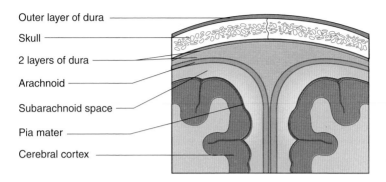

Outer layer of dura
Skull
2 layers of dura
Arachnoid
Subarachnoid space
Pia mater
Cerebral cortex

Fig. 5.16   Diagram of the meninges.

The inner layer of dura mater also encloses the spinal cord and continues down as far as the sacrum.

The *subdural space* is a potential rather than an actual space lying between the dura mater and the arachnoid mater.

The arachnoid mater is a delicate membrane lying immediately beneath the dura and dips down with it between the main portions of the brain.

The *subarachnoid space* lies between the arachnoid mater and pia mater and is filled with cerebrospinal fluid. Between the cerebellum and the medulla oblongata there is a comparatively large space, the *cisterna magna*. This can be used to obtain a sample of cerebrospinal fluid in small children. The arachnoid mater accompanies the dura as a covering to the spinal cord and extends to the sacrum.

The *pia mater* is a thin, vascular membrane which is in contact with the surface of the brain and spinal cord and which dips into all the convolutions. Inflammation of the meninges is called meningitis and results from viral and bacterial infections.

The cerebrospinal fluid

The cerebrospinal fluid is a clear, colourless liquid filling the subarachnoid space and the ventricles of the brain. It is secreted by the *choroid plexuses* in the ventricles, and passes from the two lateral ventricles, which communicate with each other and with the third ventricle through the interventricular foramen, to the third ventricle and then through a narrow tube, called the *aqueduct*, into the fourth ventricle. There are three small openings in the roof of the fourth ventricle through which the cerebrospinal fluid passes into the subarachnoid space in which it circulates around the outside of the brain and spinal cord. It is finally absorbed through the *arachnoid granulations*, which are small projections of arachnoid mater, into the venous sinuses.

The cerebrospinal fluid is similar to blood plasma in composition although it has only a small amount of protein. There are about 120 mL of fluid altogether with a pressure of 60–150 mm of water. It usually contains 200–300 mg protein per litre and about 2.8–4.4 mmol glucose per litre. These amounts may be altered in disease.

The main function of the cerebrospinal fluid is to protect the brain and spinal cord by forming a liquid cushion between the delicate nerve tissues and the walls of the bony cavities in which they lie. The cerebrospinal fluid enables the pressure within the skull to be kept constant and carries away waste and toxic substances.

PERIPHERAL NERVOUS SYSTEM

The cranial nerves

The cranial nerves originate or terminate within the brain. Some are motor nerves, some are sensory nerves, and some are mixed nerves (Table 5.1).

The spinal nerves

The spinal nerves are divided into groups according to the region of the cord from which they arise (Fig. 5.17). There are:

- Eight pairs of cervical nerves, one above the atlas and one below each of the cervical vertebrae
- Twelve pairs of thoracic nerves, one below each thoracic vertebra
- Five pairs of lumbar nerves ⎫
- Five pairs of sacral nerves ⎬ derived from the cauda equina
- One pair of coccygeal nerves. ⎭

The spinal nerves give off short posterior branches which supply the muscles of the back of the neck and trunk, and long anterior branches, which provide the nerves of the limbs and the sides and front of the trunk.

In certain regions these nerves branch immediately they leave the spinal canal, and the branches join up with one another to form the nerves supplying the various muscles and parts. This interbranching

Table 5.1   Cranial nerves

Name		Type	Function and distribution
I.	Olfactory nerve	Sensory	The nerve of smell. Starts in the nose and passes to the olfactory bulb
II.	Optic nerve	Sensory	The nerve of sight. Starts in the retina and passes to the lateral geniculate body
III.	Oculomotor nerve	Motor	Arises in the midbrain and ends in the muscles which move the eye
IV.	Trochlear nerve	Motor	As the third cranial nerve
V.	Trigeminal nerve	Motor and sensory	Supplies the muscles of mastication and has three sensory branches – ophthalmic, maxillary and mandibular
VI.	Abducent nerve	Motor	Arises in the pons and ends in one of the muscles moving the eye
VII.	Facial nerve	Motor and sensory	Supplies the muscles of facial expression and is sensory from the tongue
VIII.	Vestibulocochlear nerve	Sensory	Branches from the ear and the cochlea give hearing and the sense of balance
IX.	Glossopharyngeal nerve	Motor and sensory	The nerve of taste. Also sends motor fibres to the pharynx
X.	Vagus nerve	Motor and sensory	Supplies the digestive tract controlling both secretion and movement
XI.	Accessory nerve	Motor	Supplies the muscles of the neck, most of the pharynx and soft palate
XII.	Hypoglossal	Motor	Supplies the tongue

is called a *plexus*. Plexuses are formed in all regions except the thoracic region.

The cervical nerves form two plexuses:

1. The *cervical plexus*, which supplies muscles of the neck and shoulder, and also gives off the phrenic nerve supplying the diaphragm.
2. The *brachial plexus*, which supplies the upper limb.

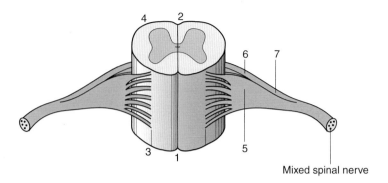

Fig. 5.17 A section of the spinal cord, showing a pair of spinal nerves arising from it. (1, 2) Fissures dividing the right and left sides of the cord. (3, 4) Smaller fissures at the side from which the nerves arise. (5) The anterior root of the spinal nerve. (6) The posterior or sensory root, on which the posterior root ganglion (7) shows as a swelling close to the point where the two roots join.

The brachial plexus gives off three main nerves: the radial; median; and ulnar nerves (Fig. 5.18). The radial nerve runs round the back of the humerus and down the outer side of the forearm. It supplies the extensor muscles of the elbow, wrist and hand. It is exposed to pressure in the armpit and against the humerus, and injury to it is the cause of wrist-drop, when the joint is flexed and cannot be extended. This may result from the use of badly padded, cheap crutches, with no hand rest, pressing in the armpit, or from pressure by the edge of the operating table on the nerve against the humerus, if the patient's arms are allowed to hang down during an operation.

The ulnar and median nerves run down the inner side and middle of the limb, respectively, and supply the flexor muscles of the wrist and hand. Injury to them causes hyperextension and gives rise to the 'claw-like' hand, the unopposed extensor muscles coming into play. A fourth smaller nerve, the *musculocutaneous nerve*, supplies the flexors of the elbow joint, the biceps and brachialis muscles. It is the ulnar nerve, crossing in the groove between the back surface of the internal epicondyle of the humerus and the olecranon, which is knocked when we say we have 'knocked our funny bone', the tingling pain passing down the nerve to the hand.

The *thoracic nerves* supply the muscles of the chest and the main part of the abdominal wall.

The *lumbar nerves* form the lumbar plexus, which gives off one main nerve, the *femoral nerve* (Fig. 5.19). This runs down beside the

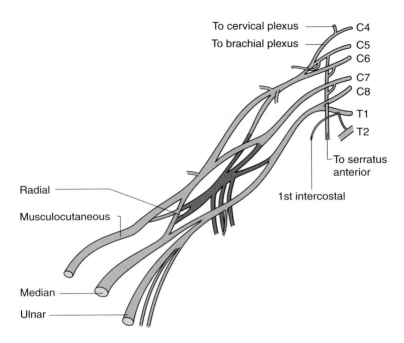

To cervical plexus — C4
To brachial plexus — C5
C6
C7
C8
T1
T2
To serratus anterior
Radial
1st intercostal
Musculocutaneous
Median
Ulnar

Fig. 5.18 The brachial plexus and the main nerves arising from it.

psoas muscle under the inguinal ligament into the front of the thigh and supplies the muscles there. The lumbar plexus also gives off branches to the lower plexus wall.

The *sacral nerves* with branches from the fourth and fifth lumbar nerves form the sacral plexus, which gives off one large nerve, the *sciatic nerve* (Fig. 5.20). This is the largest nerve in the body. It leaves the pelvis by the sciatic notch, runs across the back of the hip joint and down the back of the thigh, supplying the muscles there. It divides above the knee into two main branches:

1. The *peroneal nerve*, which supplies the muscles of the front of the leg and foot.
2. The *tibial nerve*, which supplies the muscles of the back of the leg.

The sciatic nerve therefore supplies the whole of the leg below the knee except for a small sensory branch from the femoral nerve.

The *coccygeal nerves*, with branches from the lower sacral nerves, form a second small plexus on the back of the pelvic cavity, supplying the muscles and skin in that area, for example, the muscles of

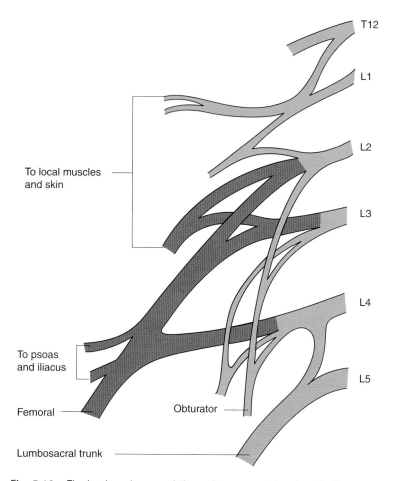

Fig. 5.19 The lumbar plexus and the main nerves arising from it. The lumbosacral trunk joins the sacral plexus.

the perineal body, the external sphincter of the anus, the skin, and other tissues of the external genitals and perineum, etc.

Both the sacral and coccygeal nerves also give off branches to the sympathetic ganglia in the pelvic area.

The autonomic nervous system

The autonomic nervous system supplies nerves to all the internal organs of the body and the blood vessels (Fig. 5.21). It is so named because these organs are self-controlled (auto = self) and not under

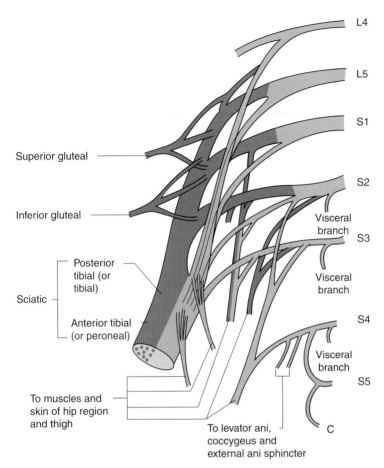

Fig. 5.20   The sacral plexus, showing the main nerves arising from it.

the control of the will. The functioning of the internal organs normally takes place without any conscious knowledge. The will does not normally affect them but the emotions do. They are affected by the hypothalamus.

The autonomic nervous system is for the most part efferent. It is made up largely of efferent neurones, both motor, supplying the involuntary muscles of the walls of organs such as the stomach, intestines, bladder, heart and blood vessels, and secretory, supplying the glands such as the liver, pancreas and kidney. There are some afferent fibres but they are comparatively few in number as the internal organs are almost insensitive. As a result, disease may attack and

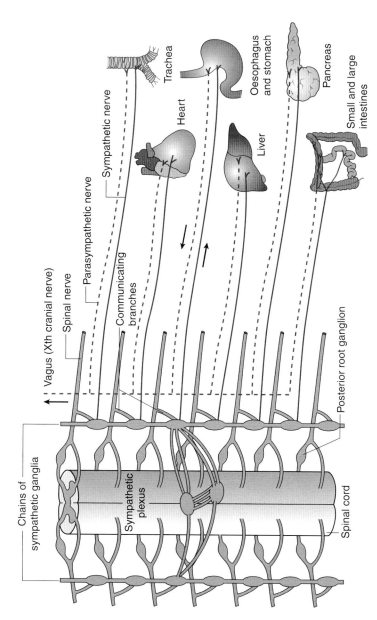

Fig. 5.21 Diagrammatic representation of the parts of the autonomic nervous system. The sympathetic nervous system is shown by the continuous lines, and the parasympathetic nervous system by broken lines.

destroy them without causing pain, and the pain that does occur is mostly due to inflammation of the lining membranes of the cavity in which they lie; for example, tuberculosis or pneumonia may affect the lung tissue without any pain but as soon as the pleura are involved, sharp pain is felt. In the same way, it hurts to cut through the abdominal wall but when a piece of bowel has been brought out of the abdomen it can be cut without causing the patient any pain, as the nurse may see in cases of colostomy.

The autonomic nervous system consists of two parts:

- The sympathetic nervous system
- The parasympathetic nervous system.

The *sympathetic nervous system* consists of a double chain of ganglia running down the trunk just in front of the vertebral column in the cervical, thoracic and lumbar regions. These ganglia are linked to one another by nerves. They receive nerves from the thoracic and upper lumbar regions of the spinal cord (Fig. 5.22). They give off

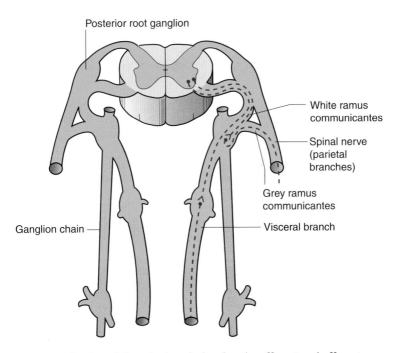

Fig. 5.22 Section of the spinal cord, showing the efferent and afferent pathways of the sympathetic nerves and the link with the spinal nerve.

nerves supplying the internal organs, the *visceral branches*, and nerves running back to the spinal nerves, the *parietal branches*. These nerves supply the blood vessels, the sweat and sebaceous glands, and the muscles that raise the hairs of the skin making them 'stand on end'. In certain regions where there are many organs requiring a nerve supply, there are additional ganglia between the two chains, linked by nerves with the chains and with one another, giving off nerves to the neighbouring organs; these are called plexuses, for example, the cardiac plexus lies behind the heart in the thoracic region, and the solar plexus lies just below the diaphragm, where the stomach, liver, kidneys, spleen and pancreas are all found.

The *parasympathetic nervous system* consists chiefly of the vagus nerve, which gives off branches to all the organs of the thorax and abdomen, but does also include branches from other cranial nerves (the third, seventh and ninth), and nerves from ganglia in the sacral region of the vertebral column.

The functions of the autonomic system

All the internal organs have, therefore, a double nerve supply, sympathetic and parasympathetic, and the two sets of nerves have, in each case, opposite actions, one stimulating and the other checking the activity of the organ. The arrangement is similar to a car, with an accelerator pedal to make it move faster and a brake pedal to check it.

The sympathetic nerves have a stimulating and quickening effect on the heart and on the respiratory system but a checking effect on digestion. They improve the circulation and cause dilation of the bronchial tubes, increasing the air intake, but they stop the secretion of digestive juices from the salivary glands and throughout the alimentary canal, and they check peristaltic action in its wall. These nerves are stimulated by strong emotions such as fear, anger and excitement. (It is because of this effect of the emotions that they are called sympathetic.) Their functions are thus closely linked to those of the adrenal medulla, which they stimulate. They help to enable the body to respond to emotion since they provide the muscles with a better supply of blood, which is rich in oxygen. This enables the individual to run away when frightened or to fight when angry, the instinctive response to these emotions. On the other hand, they are responsible for the arrest of food digestion when strong emotions are experienced, and thus may produce vomiting and emptying of the bowel, as the organs get rid of the contents, which they cannot cope with for the time being.

The parasympathetic nerves have exactly opposite effects, stimulating the digestive system and producing both a copious flow of digestive juices and peristaltic action. On the other hand, the vagus nerve slows the heart, reducing the circulation, and has a checking effect on the respiratory system, contracting the bronchial tubes. These nerves are stimulated by pleasant emotions. As a result, happiness and a contented mind tend to improve digestion. Pavlov was able to show this in a dog with a gastric fistula. When the dog was shown a bone, which pleased him, gastric juice began to pour into the stomach by reflex action through the vagus nerve. The bringing of a cat into the room angered the dog and the flow was checked, the sympathetic nerves being stimulated. Pavlov also noticed that if a bell was rung before the dog's food was brought in each day, after a time the ringing of the bell without bringing the food would cause a flow of gastric juice. This is a conditioned reflex. The animal had learned to associate the two things, so that either would produce the same reflex response. It is the same with human beings. The ringing of a dinner bell will cause secretion of digestive juices, just as will the sensation of the smell and sight of pleasant, well-served food. It is therefore of special importance to serve foods appetizingly, and to choose dishes which will give the patient as much pleasure as possible.

REFLEX ACTIONS

A reflex action is the result of the stimulation of the motor cells by stimuli brought in by afferent neurones from the tissues. Incoming stimuli can therefore, in addition to causing sensation, give rise to action; they only produce a sensation if they are passed on to the sensory centres of the brain. On the other hand, in the spinal cord and the brain they may stimulate the motor cells and give rise to an action – reflex action (Fig. 5.23). Sensory stimuli are pouring into the spinal cord and the brain from the tissues all the time. If they reach the sensory centres of the cortex of the cerebrum and stimulate them, they produce sensations of which we are conscious. If they stimulate motor cells, they produce a reflex action, for example, the touch of something hot on the skin causes immediate withdrawal of the part; a tap on the patella ligament causes contraction of the quadriceps extensor muscle and produces the 'knee jerk'; stroking of the sole of the foot causes the toes to be drawn down. In the case of the first action it may be claimed that because the heat causes an unpleasant sensation, we want to withdraw the part and the action

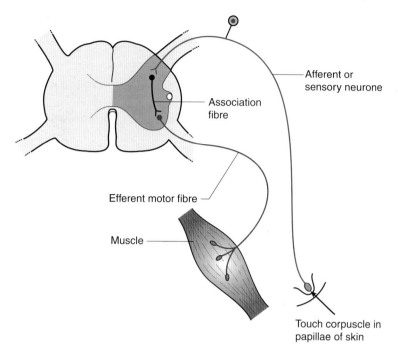

Afferent or sensory neurone

Association fibre

Efferent motor fibre

Muscle

Touch corpuscle in papillae of skin

Fig. 5.23 A reflex arc.

is voluntary. On the other hand, the action occurs in an animal when the spinal cord has been cut in the neck so that no stimuli can reach the brain and is, in fact, more marked than in an uninjured animal. The sensory stimulus is brought into the spinal cord by sensory fibres, transmitted by connector fibres to the motor cells of the anterior horn, and passed out by the motor fibres to the muscles. In the case of the knee jerk the sensory stimulus is carried into the lumbar region of the spinal cord by afferent fibres and stimulates the motor cells, which control the quadriceps.

The reason for the reflex being more marked when the spinal cord is cut off from the brain is that the cerebral centres have an inhibiting effect on a reflex action. A reflex action is not produced by the will; it is a mere response to the environment and is the only type of action found in the lower forms of life. With the development of the cerebrum, control of a reflex action and the partial replacement of reflex action by a voluntary action takes place. If we touch something hot the natural reflex is to withdraw, as the inexperienced puppy does when it touches a hot cinder in the grate. On the other hand, if we pick up a hot plate with our dinner on it, appreciating the value of the plate and the dinner, we can continue to hold the plate and put it safely down at the expense of burnt fingers. The reflex is inhibited. If we have been warned and expect the plate to be hot, this control is more easily established. In the case of the pure reflex, where the action produced has no value, as in the knee jerk and the drawing down of the toes on tickling the sole of the foot, there is still some inhibiting effect from the brain, and if the nerve path from the brain to muscle is destroyed by injury or disease above the anterior horn cell the action is much more marked.

Actions which are in the first place voluntary become in a sense reflex, for example, standing is in the first place a voluntary act which is carried out by the exercise of the will. When we have learnt to keep our balance on two feet we learn to do it by the sensations from the skin of our feet, and from muscles and joints, and the sensory organs of balance, and we can stand without voluntary effort unless disease affects the sensory nerves. In the same way, when someone learns to knit, it is at first a voluntary action. Gradually, fingers and arms learn the feel of the wool, needles and actions, and it is possible to knit without attention unless sensation is lost or a new complicated pattern is being learned.

Reflex actions may occur at three different levels in the nervous system:

1. The spinal reflex, e.g. the knee jerk
2. Reflexes occurring at the base of the brain, e.g. sneezing, coughing, vomiting, walking (cerebellar)
3. Reflexes occurring in the cerebrum and involving use of the association fibres of the brain.

Reflexes can be used to test for conditions that affect the nervous system (neurological conditions). The progression of the condition can be evaluated from these reflexes. There are deep reflexes, such as that elicited by tapping the patella tendon of the knee with a patellar hammer (the knee jerk reflex already mentioned) and superficial reflexes elicited by gently touching parts of the body such as the cornea of the eye.

PAIN

Pain is something which we all experience on occasions. Clearly, it has a useful physiological function by protecting us from harmful stimuli and warning us of disease in the body. However, pain can last well beyond the diagnosis of a disease and causes misery for many people suffering from a variety of diseases such as cancer and arthritis. Pain may be described as acute, such as the pain associated with a knock or cut, or chronic, as associated with the above diseases. A considerable effort, through the administration of prescribed pain-relieving drugs, called analgesics, or other non-pharmacological strategies forms part of nursing, and some understanding of the underlying mechanisms of pain is required by the nurse.

Pain sensation and transmission are well described by the gate control theory of pain as outlined in Fig. 5.24. The basis for this theory is that different types of fibre are involved in pain transmission and these meet in a region of the spine called the substantia

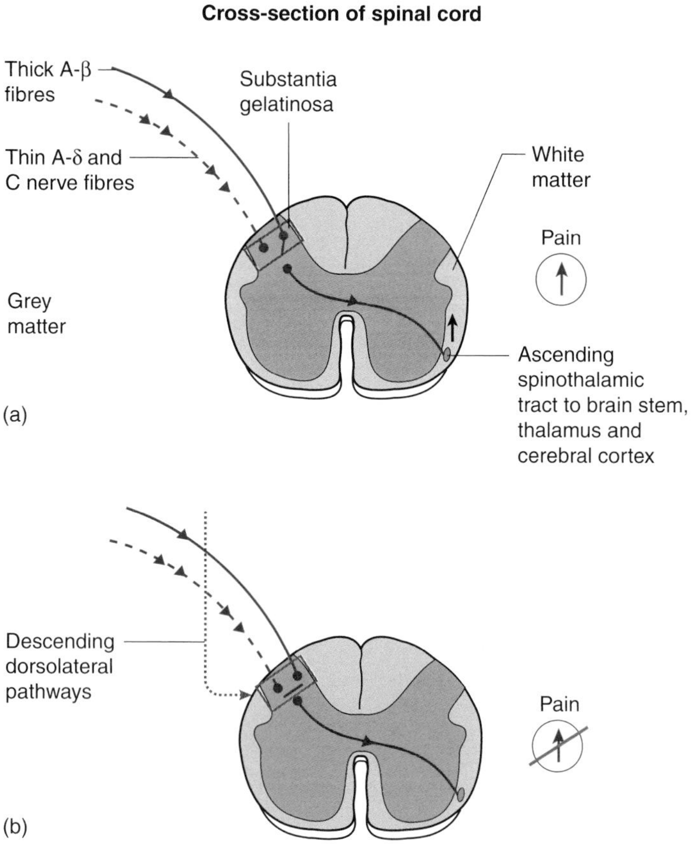

Cross-section of spinal cord

Thick A-β fibres

Substantia gelatinosa

Thin A-δ and C nerve fibres

White matter

Pain

Grey matter

(a)

Ascending spinothalamic tract to brain stem, thalamus and cerebral cortex

Descending dorsolateral pathways

Pain

(b)

Fig 5.24 The gate control theory of pain.

gelatinosa where there is a 'gate' for the transmission of pain to the areas of the brain where it is sensed. Essentially, there are thick and thin fibres and if the input to the substantia gelatinosa from the thin nerve fibres is greater than the input from the thick fibres then pain is experienced (the 'gate' is open). Alternatively, when the input from the thick fibres is greater than the input from the thin fibres then pain is not experienced (the 'gate' is closed). The 'gate' can also be closed by descending pathways from the brain and this possibly explains the actions of analgesic drugs, which work via their action on the brain and also the effect of some alternative therapies such as acupuncture, which may have the ability to release chemicals in the brain (enkephalins and endorphins), which act as natural opiates.

Sensory receptors for pain, nociceptors, exist in many tissues of the body and they respond to mechanical, chemical, temperature and inflammatory changes in the body. Pain receptors send their signals to the spinal cord, from where the signals are transmitted to the brain and sensed in the cortex. The way in which pain is experienced is very subjective and is altered by the state of the individual. For example, anxiety can increase the sensation of pain but there are also cultural differences in the experience of pain; also, the experience of associating pain with certain situations, such as visiting the dentist, can increase the sensation of pain. The experience of pain can last beyond the application of a stimulus and this may play a part in the phenomenon of chronic pain. Neurones in the spinal cord may become very sensitive to painful stimuli, thereby heightening and prolonging the experience of pain.

Pain has several physiological effects on the body, such as increasing heart rate and stomach activity. It also has an effect on the endocrine system by increasing antidiuretic and aldosterone activity, for example, which leads to water retention and increased blood pressure. Epinephrine activity is increased, thereby heightening consciousness, and increase in cortisol activity increases blood glucose levels.

SUMMARY

In this chapter the major structural features of the nervous system have been covered. The nervous system is divided into the central and peripheral systems, and the subdivisions of the peripheral system are the cranial and spinal nerves and the autonomic nervous system. Some key features of function, such as the generation and transmission of nerve impulses and the reflex arc, have also been covered.

SELF-TEST QUESTIONS

- What is an axon?
- What do the nodes of Ranvier do?
- What are the cranial nerves?
- Name the groups of spinal nerves.
- Describe a reflex arc.

Chapter 6

The ear

LEARNING OBJECTIVES

After reading this chapter you should understand:

- the structure of the ear
- the function of the ear
- the mechanism of hearing

The ear is the organ of hearing and also plays an important part in the maintenance of balance. The external ear, the middle ear and the cochlea of the internal ear are concerned with hearing; the semicircular canals, the utricle and the saccule of the internal ear are concerned with balance.

THE STRUCTURE OF THE EAR

The *external ear* has two parts: the *auricle* and the *external acoustic meatus* (Fig. 6.1). The auricle projects from the side of the head. It is composed of a thin piece of elastic fibrocartilage, covered with skin, which funnels sound waves towards the external acoustic meatus. The external acoustic meatus is a tubular passage about 4 cm long leading into the temporal bone. The outer one-third has walls of cartilage and the inner two-thirds walls of bone; the canal is curved,

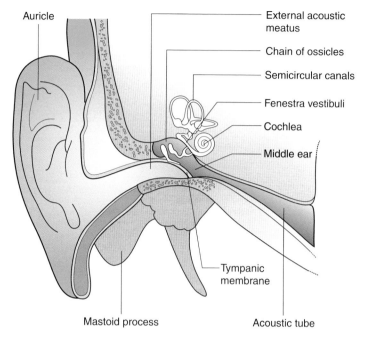

Fig. 6.1   The ear.

running first forwards and upwards, then backwards and upwards and finally forwards and slightly downwards. These curves may be straightened by gentle traction on the auricle: in adults the pull is upwards and backwards; in children backwards only; and in infants downwards and backwards. The inner end of the meatus is closed by the *tympanic membrane*. The skin lining the cartilaginous meatus contains hair follicles and numerous glands, which secrete *cerumen*. These protect the canal from foreign bodies by entangling dust and other particles but the cerumen may itself block the canal if it accumulates and will then require removal.

The *middle ear* is a small space within the temporal bone. The tympanic membrane separates it from the external ear and its further (medial) wall is formed by the lateral wall of the internal ear. The cavity is lined with mucous membrane and is filled with air, which enters from the pharynx through the acoustic tube. This equalizes air pressure on both sides of the tympanic membrane. It contains a chain of three tiny bones, the *ossicles*, which transmit the vibrations of the tympanic membrane across to the internal ear. The tympanic membrane is thin and semitransparent and the handle of

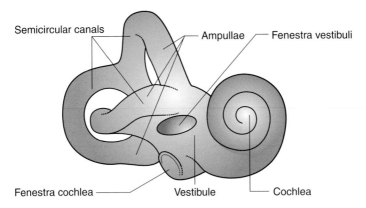

Fig. 6.2 The bony labyrinth.

the *malleus*, the first of the ossicles, is firmly attached to the inner surface. The *incus* articulates with the malleus and with the *stapes*, the base of which is attached to the *fenestra vestibuli*, which leads to the internal ear. The posterior wall of the middle ear has an irregular opening leading into the *mastoid antrum* and this in turn leads to a number of mastoid air cells. These are air-filled cavities within the bone which, like the nasal sinuses, may become infected.

The *internal ear* lies in the petrous part of the temporal bone. It consists of two parts: the bony labyrinth and the membranous labyrinth.

The bony labyrinth is again divided into three parts: the *vestibule*; the *cochlea*; and the *semicircular canals* (Fig. 6.2).

The vestibule adjoins the middle ear through two openings: the fenestra vestibuli, which is filled by the base of the stapes; and the fenestra cochlea, which is filled by fibrous tissue. At the back are openings into the semicircular canals and at the front an opening into the cochlea.

The cochlea (Fig. 6.3) is concerned with hearing. It is a spiral tube which makes two-and-three-quarter turns round a central pillar of bone called the *modiolus*. The tube is divided lengthwise into three separate tunnels by two membranes – the basilar membrane and the vestibular membrane – which stretch from the modiolus to the outer wall. The outer tunnels are the scala vestibuli above and the scala tympani below. These tunnels are filled with perilymph and they join at the top of the modiolus. The lower end of the scala tympani is closed by the fibrous fenestra cochlea. The middle tunnel is called the cochlear duct and is filled with endolymph. It is the same shape

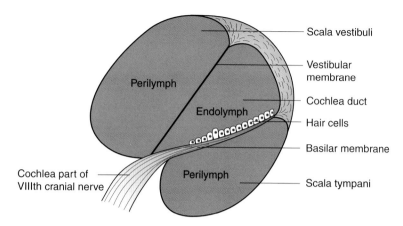

Fig. 6.3   Section through the cochlea.

as the bony labyrinth and is called the membranous labyrinth. Within the cochlear duct are the special nerve endings of the acoustic nerve called hair cells.

There are three semicircular canals. They lie above and behind the vestibule in three different planes of space: one vertical; one horizontal; and one transverse. They contain perilymph. When the position of the head is altered the movement of the endolymph stimulates special cells with hair-like processes situated at the end of each canal. This information helps in the maintenance of posture, though in daylight it is mainly the responsibility of the eyes to supply information about the position of the head in space. Overstimulation of the fluid in the semicircular canals causes giddiness.

The *membranous labyrinth* is contained within the bony labyrinth although it is much smaller. It includes the utricle, the saccule, the semicircular ducts and the cochlear duct.

The *utricle* and *saccule* are two small sacs in the vestibule which communicate with each other through a connecting tube. They contain patches of sensory hair cells which are stimulated by the action of gravity on small crystals (otoliths) which adhere to them.

The *semicircular ducts* are similar in shape to the semicircular canals and lie within them but are only a quarter of the diameter. They contain endolymph.

The *cochlear duct* is a spiral tube within the bony canal of the cochlea and lying along its outer wall. Its roof is formed by the vestibular membrane, its floor by the basilar membrane and its outer wall by the bony wall of the cochlea.

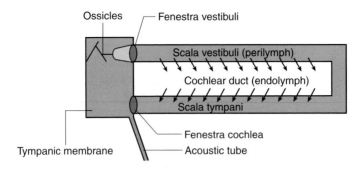

Fig. 6.4 Diagram to show how vibrations pass through the inner ear.

THE MECHANISM OF HEARING

A sound wave is a wave of *vibration* of the air set up by the vibration of an object. The vibration of a violin string or the vocal cord, for instance, sets up vibration of the air in contact with it and produces waves of vibration which spread out in all directions, like the ripples created in a still pond when a pebble is thrown in.

To produce sound, vibrations must be within certain rates. The human ear is stimulated only by vibrations at rates between 30 and 30 000 per second. The slow vibrations produce low notes and the quick vibrations produce high ones. It is for this reason that a man's voice is usually lower than a woman's as his vocal cords are longer and vibrate more slowly whereas the woman's are shorter and vibrate more quickly. It is the rapid growth in the larynx and the consequent lengthening of the boy's vocal cords at puberty that causes the breaking of the voice.

Sound waves travel at a speed of 340 metres per second. They travel much more slowly than light rays, hence the flash of lightning being seen before thunder is heard, and the farther away the storm the longer the interval between the two.

Sound waves are normally carried by the air but they also pass through solid bodies; in fact a solid carries sound more readily than air. Thus, when putting an ear to the ground, it is possible to hear footsteps at a greater distance than through the air. It is, however, normally air with which the ear is in contact.

Hearing is due to sound waves making the tympanic membrane vibrate; this sets the ossicles and fenestra vestibuli vibrating, which in turn causes vibration of the perilymph (Fig. 6.4). As fluid is incompressible, the perilymph can vibrate only if the fenestra cochlea is able to bulge outwards as the fenestra vestibuli bulges inwards. Hence the need for two windows in the internal ear. Vibration of the perilymph gives rise to vibration of the endolymph, and this affects the little hairs which jut into it and stimulates the endings of the vestibulocochlear nerve in the membranous cochlea. The nerve carries the stimulus to the centre of hearing in the temporal lobe of the brain, where it is appreciated and interpreted.

The appreciation of sound will result from any stimulus brought by the auditory nerve to the centre of hearing, but the meaning given to the sound will depend on previous experience and the power of reasoning.

SELF-TEST QUESTIONS

- What function do the semicircular canals serve?
- What are the ossicles of the ear?

Chapter 7

The eye

LEARNING OBJECTIVES

After reading this chapter you should understand:

- the structure of the eye
- the mechanism of sight
- the protection of the eye

Situated in the orbit, the eye is the organ of sight and, as such, its main function is to focus light onto the retina. The retina is composed of nervous tissue which sends signals generated by light to the brain. This chapter will look at how the structure of the eye can achieve this and how the eye, which is a very sensitive organ, is protected in order to maintain its function.

THE STRUCTURE OF THE EYE

The eye is spherical and is embedded in fat. It has three coats: an outer fibrous coat; a vascular, pigmented coat; and an inner nervous coat (Fig. 7.1).

The outer fibrous coat has two parts. The posterior part is opaque and is called the *sclera*; it is a firm membrane, which preserves the

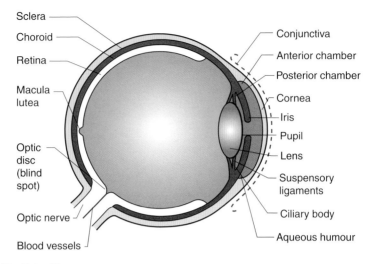

Fig. 7.1   The eye.

shape of the eyeball. Its external surface is white and forms the white of the eye. The anterior part of the sclera is covered with *conjunctiva*, which is reflected onto it from the inner side of the eyelid and is continuous with the corneal epithelium covering the cornea. The *cornea* is the anterior part of the fibrous coat. It projects a little from the surface of the eye and is transparent, allowing light rays to enter the eye and bending them to focus on the retina (refraction).

The vascular, pigmented coat has three parts. The *choroid* lines all but the front part of the eye. It is dark brown and supplies blood to the other layers of the eye, particularly the retina. The ciliary body is a thickened part of the middle coat containing muscular and glandular tissue. The *ciliary muscles* control the shape of the lens, enabling it to focus light rays from near or far away as required. They are known as the muscles of accommodation. The ciliary glands produce a watery fluid, the *aqueous humour*, which fills the eye in front of the lens and passes into the veins through small openings in the angle between the iris and the cornea. The iris is the coloured part of the eye. It lies between the cornea and the lens and divides the space between them into anterior and posterior chambers. The *iris* contains muscular tissue arranged in circular and radiating fibres; the circular fibres contract the pupil and the radiating fibres dilate it. There is a circular opening in the centre called the *pupil* which is contracted in bright light to prevent too much light entering the eye and dilated in poor light to allow as much light as possible to reach the retina (Fig. 7.2).

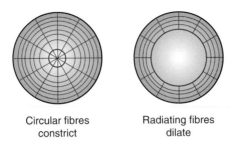

Circular fibres
constrict

Radiating fibres
dilate

Fig. 7.2   The pupil.

THE EYE

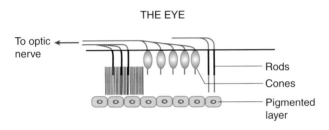

To optic
nerve

Rods

Cones

Pigmented
layer

Fig. 7.3   Layer of rods and cones on the outside of the retina.

The inner lining of the eye is called the *retina*. It is a delicate membrane adapted for the reception of light rays and contains many nerve cells and fibres. It is made of *rods* and *cones* (Fig. 7.3), which are thought to have separate functions. The cones are more numerous in the centre of the retina and are responsible for detailed vision and colour perception; the rods are more numerous around the outer edge of the retina and are sensitive to the movements of objects within the field of vision. They contain a pigment called visual purple for the synthesis of which vitamin A is required; lack of vitamin A in the diet may result in night blindness. Near the centre of the back of the retina there is an oval yellowish area called the *macula lutea* where only cones are present and which is the area where vision is most perfect. About 3 mm to the nasal side of the macula lutea the optic nerve leaves the eye; this area is called the optic disc and, because it is insensitive to light, it is called the *blind spot*. An object can only be opposite the blind spot in one eye at a time. To find a blind spot, mark a piece of paper as shown here:

X   •

Now shut the left eye, fix the right eye steadily on the cross, and move the paper slowly backwards and forwards at eye level.

At a certain point the dot will disappear because it is opposite the blind spot.

The contents of the eye are the:

- Aqueous humour
- Vitreous humour
- Lens.

The aqueous humour has already been described.

The *vitreous humour* is a colourless, transparent, jelly-like substance which maintains the shape of the eyeball.

The *lens* is situated immediately behind the iris. It is a transparent, biconvex body enclosed in a transparent, elastic capsule from which ligaments pass to the ciliary body. These *suspensory ligaments* hold the lens in position and are the means by which the ciliary muscles exert their pull on the lens, altering its shape for near or far vision.

THE MECHANISM OF SIGHT

As light rays pass through the transparent cornea, aqueous humour and lens, they are bent by *refraction*. This makes it possible for light from a large area to be focused on a small area of the retina. Parallel light rays striking a convex lens are bent towards a focal point on the retina (Fig. 7.4). If the object is less than 7 m away the curvature of the lens must be increased to enable the focus to be on the retina. This is called *accommodation* (Fig. 7.5). Far vision can be achieved with the lens in its normal resting position.

Some people are naturally *short-sighted*: their eyes are too long and so the retina is farther from the lens than it should be, therefore the focusing point lies in front of it. In this case the near object can be seen with the flatter lens of the eye at rest (normally used for far sight) but for far sight, concave glasses are necessary to throw the focus point farther back.

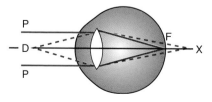

Fig. 7.4 The normal eye at rest. P, Parallel rays from a distant object are focused by the lens on the retina (F). Dotted lines show the diverging rays from a near object (D) of which the focus falls behind the retina at X.

Other people are naturally *long-sighted*: their eyes are too short and so the retina is too close to the lens, therefore the focusing point lies behind it. In this case the far object can be seen by using the thicker, more curved lens used by the normal person for near sight, as this will bend the light rays more and bring the focusing point forward; for near sight, since they are already using the more curved lens, they must be provided with convex glasses to bend the light rays from the near object still more acutely (Figs 7.6 and 7.7).

The eye is moved within the orbit by six *orbital muscles*, which are ribbon-like and attached to the sclera. These muscles pull on the eyes and co-ordinate their movement so that both eyes focus on the same object. A weakness in one or more muscles will cause one eye to deviate, a condition that is commonly known as a squint.

PROTECTION OF THE EYES

The eyes are very delicate organs and are protected by the eyebrows, the eyelids and the lacrimal apparatus, as well as by the bony orbits in which they lie embedded in fatty tissue.

The overhanging brows protect them from injury and excessive light, while the hairs entangle sweat and prevent it from running into the eyes.

The *eyelids* consist of a plate of fibrous tissue covered by skin and lined with mucous membrane. The edges of the lids are provided with

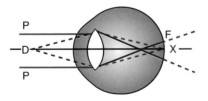

Fig. 7.5   The normal eye during accommodation. Rays (dotted lines) from a near object (D) are focused on the retina at X by the more curved lens. P, Parallel rays from a distant object are now focused in front of the retina (F).

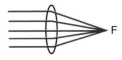

Fig. 7.6   A convex lens, showing how parallel light rays are brought to focus at F.

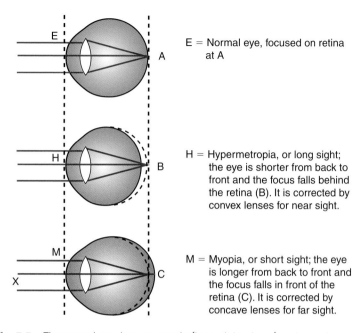

E = Normal eye, focused on retina at A

H = Hypermetropia, or long sight; the eye is shorter from back to front and the focus falls behind the retina (B). It is corrected by convex lenses for near sight.

M = Myopia, or short sight; the eye is longer from back to front and the focus falls in front of the retina (C). It is corrected by concave lenses for far sight.

Fig. 7.7 The normal eye, hypermetropia (long-sightedness) and myopia (short-sightedness).

hairs – the eyelashes – which keep out dust, insects and too much light. The transparent mucous membrane that lines the lids is reflected over the front of the eyeball and is called the *conjunctiva*. This results in the formation of upper and lower conjunctival sacs under the upper and lower lids, respectively. Dust and bacteria tend to stick to the moist surface of this membrane, and to keep it clean it is constantly washed by the *lacrimal apparatus*.

The lacrimal apparatus (Fig. 7.8) consists of:

1. The *lacrimal gland*, lying over the eye at the outer side and secreting lacrimal fluid into the conjunctival sac
2. Two fine canals, called *lacrimal canaliculi*, leading from the inner angle of the lids to the lacrimal sac
3. The *lacrimal sac*, which lies at the inner angle of the eyelids in the groove on the lacrimal bone
4. The *nasolacrimal duct*, which runs from the lacrimal sac down to the nose.

The opening into the canaliculi can be seen at the inner angle of the eyelids and is called the *punctum*.

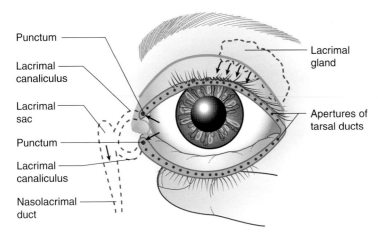

Punctum

Lacrimal
canaliculus

Lacrimal
sac

Punctum

Lacrimal
canaliculus

Nasolacrimal
duct

Lacrimal
gland

Apertures of
tarsal ducts

Fig. 7.8 The lacrimal apparatus.

The fluid secreted by the lacrimal glands washes over the eyeball and is swept up by the blinking action of the eyelids. The muscles that cause the blink press on the lacrimal sac and contract it so that as they relax the sac expands and 'sucks' the fluid from the edges of the lids along the fine canals into the sac; from this it runs by gravity down into the nose. Thus the window that admits light into the eye is constantly irrigated by a gentle stream of fluid, which keeps it clean and washes away germs and harmful substances. The fluid is composed of water, salts and an antibacterial substance called *lysozyme*.

SELF-TEST QUESTIONS

- Draw and label a diagram of the eye.
- In the retina, what is the function of the rods and cones?
- Explain long- and short-sightedness.

Chapter 8

Endocrine glands

LEARNING OBJECTIVES

After reading this chapter you should understand:

- the function of hormones
- the control of hormone secretion
- the function of specific glands

The endocrine glands are organs producing secretions called *hormones*, which are poured directly into the blood stream from the glandular cells. It is for this reason that they are also known as ductless glands. This chapter is mainly concerned with the overall control of the endocrine system, which is composed of all the endocrine glands, by the pituitary in the brain. The function of the thyroid and parathyroid glands will be considered in some detail along with the adrenal glands, although the gonads, which are also endocrine glands, are considered in Chapter 26.

Hormones are organic compounds manufactured by the glands from substances carried in the blood. They are mainly protein derivatives although some are steroids. They are then carried by the blood stream to other parts of the body where they have a specific effect.

The endocrine glands have been investigated and are described as though they are separate entities but in fact their functions are closely interrelated. Initially the functions of the hormones were deduced by observing the effects of disease, destruction or overgrowth of glands. In recent years, hormones have been isolated, obtained in pure form, analysed and, in some cases, successfully synthesized. The glands secrete their hormones continuously but the quantity of secretion can be increased or decreased according to body needs. Control of hormone secretion occurs in several different ways:

1. Nerve cells produce chemical substances that are carried to the gland and cause secretion.
2. The gland responds to impulses from the autonomic nervous system.
3. One gland produces a hormone, which affects a second gland. The second gland produces its hormone, which influences the secretion of the first gland. This is called a 'feedback' mechanism.
4. The gland responds to blood levels of substances other than hormones.

THE PITUITARY

The pituitary gland lies in the hypophyseal fossa of the sphenoid bone in the base of the skull. It is attached by a *neural stalk* to the optic chiasma at the base of the brain (Fig. 8.1).

The gland consists of an *anterior lobe*, or adenohypophysis, and a *posterior*, or neural, lobe. The anterior lobe is an endocrine gland in the true sense, while the posterior lobe is derived from the brain and consists of nervous tissue; it is connected directly to the *hypothalamus*. The two lobes are essentially two different endocrine glands and are commonly called the anterior and posterior pituitary glands.

The anterior lobe of the pituitary is sometimes referred to as the 'master gland' of the endocrine system because it has an important influence in regulating the function of other glands. However, the glands really act in concert, one becoming active if the others are not producing sufficiently and output of hormones being decreased when other glands are active.

The anterior lobe produces a number of hormones:

- Thyroid-stimulating hormone (TSH or thyrotrophic hormone)
- Adrenocorticotrophic hormone (ACTH)
- Somatotropin
- Follicle-stimulating hormone (FSH)

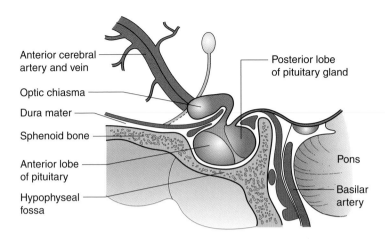

Fig. 8.1   The pituitary gland, shown in section, lying in the sella turcica or hypophyseal fossa. Note the stalk joining it to the base of the brain.

- Luteinizing hormone (LH)
- Prolactin.

1. *Thyroid-stimulating hormone* influences all aspects of thyroid gland function including stimulation of the accumulation of iodine in the gland for conversion into thyroid hormones, manufacture of the hormones, and their release into the circulatory system. Through its action on the thyroid gland, TSH is involved in the regulation of metabolic rate, breakdown of fat and the increase of the water content of certain tissues.

2. *Adrenocorticotrophic hormone* regulates the development, maintenance and secretion of the cortex of the suprarenal glands. General metabolic effects of this hormone include mobilization of fats, leading to hypoglycaemia and increase in muscle glycogen. It is also involved in body resistance to stress.

3. *Somatotrophic (growth) hormone* exerts its influence mainly on the hard tissues of the body, though there is some effect on the soft tissues. The hormone increases the rate of growth and maintains size once maturity has been reached. It controls the rate of growth in epiphyses and other ossification centres of the skeletal system. Oversecretion of this hormone causes overgrowth of the long bones in children (gigantism) and acromegaly in adults. In acromegaly the bones cannot increase in length because the epiphyseal plates have closed so they become thicker and coarser; the lower jaw, hands and feet are

particularly affected. Undersecretion of the growth hormone results in dwarfism. People who are very short or very tall, due to under- or oversecretion of this hormone, are usually of normal intelligence, unlike those who suffer from undersecretion of the thyroid gland. Metabolic effects of this hormone include an increase in the conversion of carbohydrates to amino acids and in the uptake of amino acids by the cells, mobilization of fat from the storage areas with increased fat metabolism, and an increased blood sugar level.

4. *Follicle-stimulating hormone* controls the maturation of ovarian follicles in the female and the production of sperm in the male.

5. *Luteinizing hormone* causes changes in the female which lead to the formation of the *corpus luteum*; it also helps to prepare the breasts for the secretion of milk. In the male, a comparable hormone is called *interstitial cell stimulating hormone (ICSH)*, which acts on the testes and controls the secretion of the male sex hormone testosterone.

6. *Lactogenic hormone* (prolactin) is one of several hormones involved in the production of milk by the breasts and appears to function only in females.

The posterior lobe releases two hormones:

- Oxytocin
- Antidiuretic hormone (ADH or vasopressin).

It is important to note that these hormones are produced in the hypothalamus and only stored in, and released from, the posterior lobe.

1. *Oxytocin* exerts its effects mainly on the smooth muscle of the pregnant uterus and the cells around the ducts of the breasts, although it does also promote a generalized contraction of unstriped muscle throughout the body.

2. *Antidiuretic hormone* causes an increase in the reabsorption of water by the kidney tubules so that less urine is excreted. Undersecretion of ADH causes less water to be reabsorbed and excessive amounts of very dilute urine are excreted – a condition known as diabetes insipidus. The hormone also causes a certain amount of vasoconstriction with a consequent rise in blood pressure but in human beings this occurs mainly in the coronary vessels.

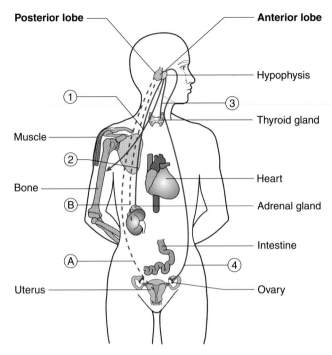

Posterior lobe — Anterior lobe — Hypophysis — Thyroid gland

Muscle — Bone — Heart — Adrenal gland — Intestine — Uterus — Ovary

Fig. 8.2   Diagram to show the effects of the pituitary hormones. The numbers and letters refer to those in Fig. 8.3.

The effects of the pituitary hormones are summarized in Figs 8.2 and 8.3.

THE THYROID GLAND

The thyroid gland is situated at the front and sides of the neck, opposite the lower cervical and first thoracic vertebrae. It consists of two lobes, one on either side, joined by a narrower portion called the isthmus, which crosses in front of the trachea just below the larynx (Fig. 8.4). The gland is formed of a number of closed follicles containing a yellow semifluid material called colloid (Fig. 8.5).

The cells produce two hormones called *thyroxine* and *tri-iodothyronine* (also known as T_4 and T_3, respectively), which may be released directly into the blood stream if it is required or may be linked to the protein substance thyroglobulin and stored in the colloid. The amino acid tyrosine and the mineral iodine are both essential for the formation of thyroid hormones. Tyrosine in the thyroid hormones

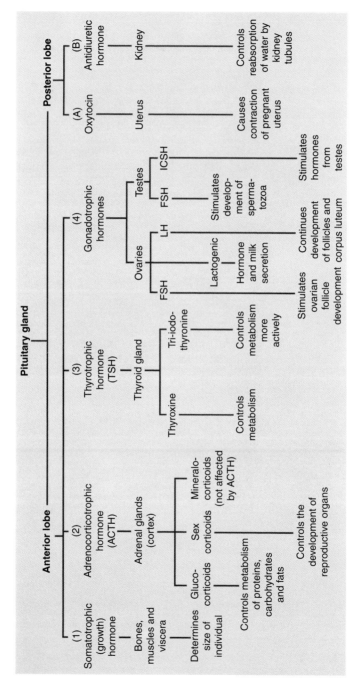

Fig. 8.3 Diagram to show the actions of the pituitary hormones. FSH, follicle-stimulating hormone; LH, luteinizing hormone; ICSH, interstitial cell stimulating hormone.

Right   **Left**

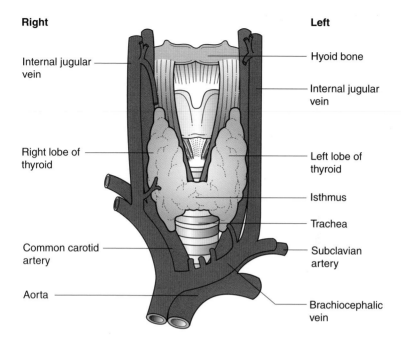

Internal jugular vein — Hyoid bone

Internal jugular vein

Right lobe of thyroid — Left lobe of thyroid

Isthmus

Trachea

Common carotid artery — Subclavian artery

Aorta — Brachiocephalic vein

Fig. 8.4 The thyroid gland.

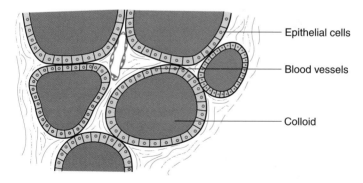

Epithelial cells

Blood vessels

Colloid

Fig. 8.5 Thyroid follicles.

binds iodide. Thyroxine binds four iodide ions (hence T_4) and tri-iodothyronine binds three iodide ions (hence T_3).

The function of thyroxine is to regulate metabolism in the tissues. Together with growth hormone, it ensures proper development of the brain; it increases urine production, protein breakdown and the uptake of glucose by the cells. Tri-iodothyronine has a more immediate, though similar, effect to thyroxine.

Undersecretion of thyroid hormones in a child produces the condition of thyroid cretinism, which, if untreated, results in a person of very small stature with learning disabilities. In an adult, undersecretion produces the condition known as *myxoedema*. In both these conditions the skin is dry and coarse, and the hair dry, coarse and lank. The metabolic rate is lowered so the patient is obese, has a low body temperature and feels the cold. Oversecretion of the hormones produces *thyrotoxicosis*, a condition in which there is an increase in the metabolic rate. The individual is anxious and nervous and has a fast pulse rate. The skin is fine and moist, the patient feels the heat and loses weight in spite of having a good appetite. Patients may or may not have protruding eyeballs, a condition known as exophthalmos.

Any enlargement of the gland is known as *goitre* and this may be present with or without hyperthyroidism. The presence of a goitre may cause pressure on the trachea or on the recurrent laryngeal nerve causing hoarseness.

THE PARATHYROID GLANDS

The parathyroid glands usually lie between the posterior borders of the lobes of the thyroid gland and its capsule. They are about the size of a pea and there are usually four: two behind each lobe. This number occasionally varies.

The glands produce a hormone called *parathormone*, which regulates the distribution and metabolism of calcium and phosphorus in the body. Oversecretion causes calcium to be moved from the bones into the blood, from where it is excreted in the urine. The bones become porous and brittle and the increased level of blood calcium may cause the formation of kidney stones (renal calculi). Undersecretion of the hormone causes low blood calcium levels, resulting in the muscular rigidity and spasm seen in the condition known as *tetany*. This complication may occasionally be seen after surgery on the thyroid gland in which the parathyroids have inadvertently been removed. Calcium is given to correct the condition.

THE ADRENAL GLANDS

The adrenal glands lie one above and one in front of the upper end of each kidney, behind the peritoneum. They are surrounded by areolar tissue containing a considerable amount of fat. Each gland consists of two quite separate endocrine glands: the outer part called the cortex; and the inner part the medulla. They secrete hormones as follows:

Cortex

- Mineralocorticoids
- Glucocorticoids
- Sex hormones.

Medulla

- Epinephrine
- Norepinephrine.

Cortex

The cortex is subdivided into three zones: the outer zone produces the mineralocorticoids; the central zone produces the glucocorticoids; and the inner zone produces sex hormones.

Mineralocorticoids are the steroids that regulate electrolyte metabolism. *Aldosterone* is an important mineralocorticoid and it has a regulatory effect on the relative concentrations of minerals in the body fluids, particularly sodium and potassium. It therefore also affects the water content of the tissues. When there is a deficiency of this hormone, there is increased excretion of sodium and chloride ions, too much water is lost from the body in the urine and there is a corresponding fall in the concentrations of sodium, chloride and bicarbonate in the blood, resulting in a lowered pH (acidosis).

Glucocorticoids are important in carbohydrate metabolism. They increase the conversion of protein to glycogen for storage by the liver (gluconeogenesis) and decrease the use of glucose by the cells, thus increasing the blood sugar level. *Cortisone, cortisol* (hydrocortisone) and *corticosterone* are the primary glucocorticoids and they may be given to people who have chronic inflammatory and allergic responses, such as in *rheumatoid arthritis*, because the steroids stimulate the anti-inflammatory and repair mechanisms of the body. Chronic stress of any kind increases production of the glucocorticoids, which in turn help the body to resist stress; however, continuing high levels of glucocorticoids may increase the likelihood of ulcer formation, increase blood pressure and lower the body's resistance to infection by causing atrophy of lymphatic tissue.

Sex hormones are the *androgens* (male hormones) and *oestrogens* (female hormones) and the effects are similar to those of the hormones produced by the testes and ovaries. Adrenal androgens and oestrogens are produced in minute quantities in both sexes and affect the development and function of the reproductive organs and the physical and temperamental characteristics of men and women. Their effects become noticeable when there is oversecretion; for example, an adrenal tumour in a woman may produce secondary masculine characteristics such as the growth of hair on the face and deepening of the voice.

Undersecretion of the cortical hormones results in a condition known as *Addison's disease*, which produces symptoms of anaemia, muscular weakness, low blood pressure, low blood sugar level, and bronzing of the skin and mucous membranes.

Oversecretion of the cortical hormones results in several disorders. *Cushing's disease* is due mainly to overproduction of the glucocorticoids. There is excess fatty tissue on the trunk and face but not the limbs; sodium is retained, so there is oedema, increased plasma volume and a raised pH (alkalosis).

Medulla

The medulla of the adrenal glands produces two hormones, *epinephrine* and *norepinephrine*, which have effects similar to those obtained by stimulation of the sympathetic nervous system. Because control of the medulla is by way of preganglionic neurones of the sympathetic nervous system, without interruption, response to stimulation is very rapid and creates a set of conditions in which the body is prepared for 'fight or flight'. Epinephrine increases the strength and rate of the heart beat; causes dilation of arteries supplying the heart and skeletal muscles but constriction of other arteries; increases rate and depth of respiration; increases blood sugar level by promoting breakdown of liver glycogen; and stimulates the general metabolic activity of the cells.

THE THYMUS GLAND

The thymus gland is an organ in the lower part of the neck and chest between the lungs over the heart. It varies in size with age and grows until 2 years of age; it then shrinks so that in the adult only fibrous remnants are found. When fully developed, it is greyish-pink and consists of two or three lobes. Its structure resembles that of a lymph node and it is similarly associated with antibody forma-tion. No hormone has yet been isolated from it and it may well belong to the circulatory rather than the endocrine system.

THE PINEAL BODY

The pineal body, also called the pineal gland, is a small reddish body about the size of a cherry stone lying behind the third ventri-cle of the brain. Its function is unknown. In later life, it becomes calc-ified and acts as a useful landmark in X-rays of the skull. The pineal body is implicated in the regulation of circadian (day/night) rhy-thms through the secretion of melatonin.

THE GONADS

The gonads, which are responsible for the production of ova in females and sperm in males, are under the control of FSH and LH. Their function, which also includes endocrine functions, is dis-cussed in Chapter 26.

THE PANCREAS

The pancreas has an endocrine function. It contains an area known as the *islets of Langerhans*, which produces *glucagon* and *insulin* from the α and β cells, respectively. These hormones influence glucose metabolism, which is discussed in Chapters 20 and 22.

SELF-TEST QUESTIONS

- What is an endocrine gland?
- How do hormones get from glands to target tissues?
- Why is the pituitary so important?
- What is the function of the parathyroid glands?
- What are the respective functions of the adrenal cortex and medulla?

SECTION 3

Posture and movement

SECTION CONTENTS

SECTION INTRODUCTION

Posture and movement, which are controlled by the central nervous system, are achieved by the combined actions of the skeletal and muscular systems. The skeletal system is composed of the bones, which make up the skeleton, and the muscular system is composed of the skeletal, or voluntary, muscles which are attached to the skeleton.

There are different types of bone, and these, in turn, are capable of different types and degrees of movement. Movement between two bones is called 'articulation' and the system of joints whereby this is achieved is described.

Skeletal muscle is a unique type of tissue and its gross structure and organization into groups of muscles, as well as the mechanism whereby movement is achieved, are all considered here. The use of chemical energy, in the form of ATP, which is produced in metabolism, is introduced here.

Chapter **9**

Development and types of bone

LEARNING OBJECTIVES

After reading this chapter you should understand:

- the formation of bone
- the role of vitamin D in bone formation
- the structure of bone
- the different types of bone
- the growth of a long bone

The skeletal system is made up of just over 200 bones, joined together to provide a strong, movable, living framework for the body.
   The skeletal system has four main functions:

1. It supports and protects the surrounding soft tissues and vital organs.

2. It assists in body movement by giving attachment to muscles and providing leverage at joints.
3. It manufactures blood cells in the red bone marrow.
4. It provides storage for mineral salts, particularly phosphorus and calcium.

This chapter looks at the developmental processes undergone by bone in the production and maintenance of the human skeleton. The ways in which different types of bone can be described anatomically are also outlined.

Cartilage provides the environment in which bone develops. Development is from spindle-shaped cells, called *osteoblasts*, which are found underneath the tough sheath of fibrous tissue that covers the bone, called the *periosteum*, and in cavities within the bone. These cells are able to convert the soluble salts of calcium and magnesium in the blood, such as calcium chloride, into insoluble calcium salts, chiefly calcium phosphate. A second set of cells is also present; these can change the insoluble calcium phosphate into soluble calcium salts, which are carried away by the blood. These cells are capable of causing the absorption of any unwanted bone and are known as *osteoclasts*. Both types of cell are active during the growth period: osteoblasts producing bone and osteoclasts removing it to maintain the form and proportion of the bone. For example, osteoblasts build up bone on the surface of a hollow bone, while osteoclasts absorb bone on the inner side to enlarge the cavity and prevent the bone from becoming too heavy.

A loss of bone density due to excessive absorption of bone material is called *osteoporosis*; it leads to a weakening of bones, making them more susceptible to fracture. Loss of bone density is a feature of ageing, especially in postmenopausal women. Mild exercise in older people can help to reduce the effects of osteoporosis.

OSSIFICATION

There are two types of ossification: intramembranous and intracartilaginous.

Intramembranous ossification

- A process in which dense connective tissue is replaced by deposits of calcium salts, forming bone.
- The bones of the skull form by this process.

Intracartilaginous ossification

- A process in which cartilaginous structures are replaced with bone.
- Most bones form by this process.

BONE GROWTH AND REPAIR

Plentiful supplies of calcium and phosphorus are essential in the diet of the pregnant and nursing mother, the growing child, and where bone repair is taking place after injury or disease. Calcium is present in milk and eggs and in green vegetables, and phosphorus in meat, fish and the yolks of eggs. *Vitamin D* is necessary so that calcium and phosphorus may be absorbed from the intestine and fully used within the body; lack of vitamin D causes *rickets* in children and *osteomalacia* in adults. Both these conditions cause soft bones, which bend under the weight of the body and various deformities of the weight-bearing bones may result. Vitamin D is found in animal fat and fish oils and in margarine to which it has been added.

The human body can also manufacture vitamin D – the action of ultraviolet rays from the sun converts ergosterol in the skin into vitamin D.

Vitamin C is important in bone growth as it plays a part in laying down collagen, which is the main constituent of connective tissue. Vitamin C is found in fresh fruit (particularly citrus types), blackcurrants, green vegetables, tomatoes and potatoes.

Growth and development of bone is also affected by exercise and rest. Exercise causes an increased blood supply to the muscles and therefore to the underlying bones. Because the blood carries necessary building materials, exercise results in increased growth, and recognition of this fact is responsible for increased attention to physical exercise in schools. The pull of developing muscles is an important factor in determining the shape of the bones, and posture may also affect shape by altering stresses on bones. The importance of rest is interesting. The bones of a child are comparatively elastic so that at the end of a day of standing and running about, there is an appreciable loss of height. During rest the bones recover their full length, so a midday rest and a long night will help growth.

A further factor controlling growth is the secretion of certain ductless glands. This was mentioned in Chapter 8.

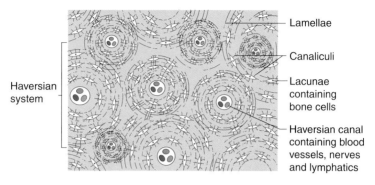

Fig. 9.1 The structure of compact bone.

TYPES OF BONE TISSUE

Bone tissue is of two types: compact and spongy.

Compact bone appears to be solid but when it is examined microscopically it is found to consist of haversian systems (Fig. 9.1), which have the following components:

- A central canal, called a *haversian canal*, which contains blood vessels, nerves and lymphatics
- Plates of bone, *lamellae*, arranged round the central canal
- Spaces called *lacunae*, between the lamellae, which contain bone cells (osteocytes) and lymph
- Fine channels, *canaliculi*, between the lacunae and the central canal, carry lymph, which brings nutrients and oxygen to the osteocytes.

Between the haversian systems, there are tiny circular plates of bone called interstitial lamellae.

Spongy bone is hard like all bone but has a spongy appearance to the naked eye. When examined under a microscope the haversian canals are seen to be much larger and there are fewer lamellae. The spaces in spongy bone are filled with red bone marrow, which consists of fat and blood cells and in which red blood cells are made.

TYPES OF BONE

There are three types of bone:

- Long bones
- Flat bones
- Irregular bones.

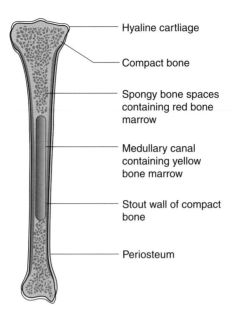

Hyaline cartliage

Compact bone

Spongy bone spaces
containing red bone
marrow

Medullary canal
containing yellow
bone marrow

Stout wall of compact
bone

Periosteum

Fig. 9.2   The structure of a long bone.

Long bones

A long bone consists of a shaft and two extremities (Fig. 9.2). The shaft has an outer layer of compact bone surrounding a central cavity called the *medullary canal* which contains yellow *bone marrow*. Yellow bone marrow, like red bone marrow, consists of fat and blood cells but it is not so rich in blood supply or in red blood cells. The extremities consist of a mass of spongy bone containing red bone marrow covered by a thin layer of compact bone. The bone is covered by a tough sheath of fibrous tissue, the periosteum; this is richly supplied with blood vessels, which pass into the bone to nourish it.

Three distinct sets of blood vessels supply the long bones:

1. Innumerable tiny arteries run into the compact bone to supply haversian canals and systems.
2. Many larger arteries pierce the compact bone to supply the spongy bone and red bone marrow. The openings by which these vessels enter the bone can be seen quite easily with the naked eye.
3. One or two large arteries supply the medullary canal. These are known as nutrient arteries and they pass through a large opening called the nutrient foramen, which runs obliquely through the shaft to the medullary canal.

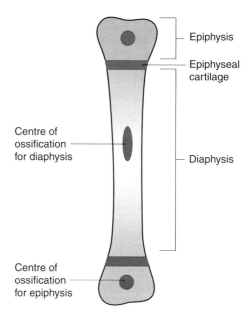

Fig. 9.3 The development of a long bone (growth in length occurs at the epiphyses).

These three sets of vessels are linked by fine branch arteries within the bone.

The periosteum nourishes the underlying bone through its blood vessels. If it is torn off the underlying bone will die, but if the bone is destroyed by disease and the periosteum remains healthy it is able to build up new bone. The periosteum is also responsible for growth in the thickness of the bone through the action of osteoblasts, adjacent to the bone surface, which are able to lay down fresh bone. The periosteum is protective in function and also provides attachment to tendons of muscles. It is not present on the joint surfaces of a bone where it is replaced by hyaline cartilage, called articular cartilage, which provides a smooth surface so that joint movements can occur without friction.

Development of long bones

Long bones develop from three *centres of ossification*, one in the shaft and one or more at either extremity (Fig. 9.3). The centre of ossification in the shaft is known as the *diaphysis*. Those at the extremities are called *epiphyses* and begin to develop after birth. From these centres

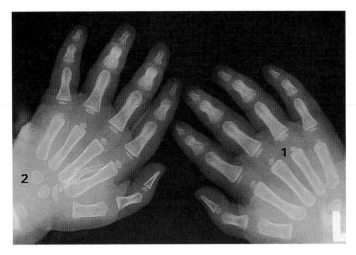

Fig. 9.4   X-ray of the hand of a 1-year-old infant. The shafts of the small long bones of the fingers and thumbs are well formed but note the gaps (1) between them and the knuckles. Note also that only two of the eight bones of the wrist (2) have begun to ossify.

ossification gradually spreads throughout the extremity (Figs 9.4 and 9.5) and is well developed by 12 years of age, though there is still a line of cartilage left between the shaft and the extremity.

The line of cartilage between epiphysis and diaphysis is known as the *epiphyseal cartilage* and it is from this cartilage that growth in length occurs. It is the shaft that grows longer and fresh bone is constantly produced at either end of it by the epiphyseal cartilage. When full growth has been attained, between the ages of 18 and 25 years, the line of cartilage turns to bone and no further growth can occur. The timing of this event varies from one bone to another and from one end of a bone to the opposite end of the same bone. For example, the lower epiphysis of the humerus unites at about the 18th year but the upper epiphysis does not join the shaft until about 2 years later.

Flat bones

Flat bones consist of two stout layers of compact bone joined by a layer of spongy bone (Fig. 9.6). These bones are also covered by periosteum from which two sets of blood vessels pass into the bone to supply the compact and spongy bone, respectively. They are found in the head, trunk, shoulder girdle and pelvic girdle, where they give protection to the delicate organs underlying them and provide

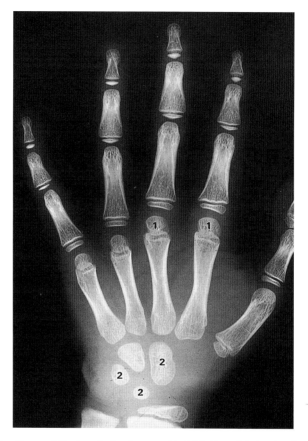

Fig. 9.5   X-ray of the hand of a 6-year-old child. Note (1) the development of the separate extremities, i.e. the epiphyses, of the long bones of the hand and (2) the ossification that has occurred in the wrist or carpal bones. Only certain bones are marked for comparison.

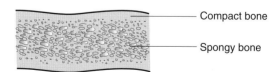

Compact bone

Spongy bone

Fig. 9.6   The structure of a flat bone.

attachment for the powerful muscles required to control the freely movable shoulder and hip joints. The layers of compact bone in the flat bones of the skull are referred to as the outer and inner tables and the spongy bone is called the *diploë*. The outer table is thick and strong

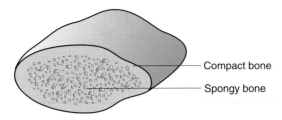

Fig. 9.7 The structure of an irregular bone.

so that it is not readily broken. In some areas the spongy bone is absorbed, leaving air-filled cavities, *sinuses*, between the two tables.

Irregular bones

Irregular bones consist of a mass of spongy bone covered by a thin layer of compact bone (Fig. 9.7). They are covered with periosteum, except on the articular surfaces, and as with flat bones, this protective covering provides two sets of blood vessels to supply both compact and spongy bone. They are found in the spinal column and the middle ear, and also at the ankle and wrist although these latter are sometimes called short bones.

SURFACE IRREGULARITIES

The surfaces of all bones are very irregular and show numbers of depressions and projections. These may be divided according to function.

Articular projections and depressions

These enter into the formation of joints and are smooth.

Articular projections

- Head – round, like a sphere or disc
- Condyle – rounded but oval in outline, like the typical knuckle bone.

Articular depressions

- Socket – a depression into which another organ fits
- Fossa – a shallow depression.

Non–articular projections and depressions

These provide attachment to muscles or ligaments and are rough.

Non-articular projections

- Process – a rough projection for muscle attachment
- Spine – a pointed rough projection
- Tuberosity – a broad rough projection
- Trochanter – a broad rough projection
- Tubercule – a small tuberosity
- Crest – a long rough narrow projecting surface.

Non-articular depressions

- Fossa – a notch in the bone
- Groove – a long narrow depression.

Other terms applied to bones

- Foramen – an opening in the bone
- Sinus – a hollow cavity in the bone.

All the rough non-articular projections provide attachment to muscles; the stronger the muscle and the more it is used, the larger and rougher the projection, providing a greater surface for muscle attachment. In a paralysed limb the processes either fail to develop, or atrophy, according to age.

Occasionally, bones get broken or fractured. However, they heal satisfactorily due to their copious blood supply. There are four recognized stages in the healing of a fractured bone:

- Formation of a haematoma (blood clot) 6–8 hours after fracture
- Formation of fibrocartilage for about 3 weeks
- Formation of bone for 3–4 months
- Remodelling.

SELF-TEST QUESTIONS

- Describe the process of ossification.
- Describe the growth of a long bone.
- Name two examples each of long, flat and irregular bones.

Chapter 10

Bones of the head and trunk

LEARNING OBJECTIVES

After reading this chapter you should understand:

- the divisions of the skeleton
- the bones of the head
- the bones of the trunk
- the structure of a vertebra
- the structure of the spine

The student is advised that it is not possible to study the bones of the human skeleton adequately without access to an entire skeleton and to disarticulated bones that can be handled and examined closely.

The skeleton (Fig. 10.1) can be divided into groups of bones:

- The bones of the head
- The bones of the trunk
- The bones of the upper limb and shoulder girdle
- The bones of the lower limb and pelvic girdle.

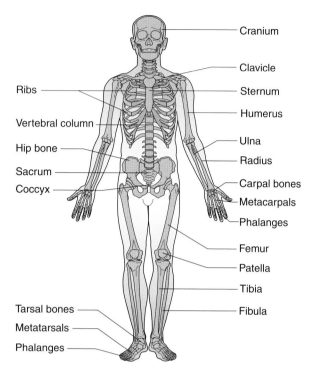

Fig. 10.1 The skeleton (male).

The bones of the head and trunk comprise the *axial skeleton*, which is the main support of the body, while the bones of the extremities are also known as the *appendicular skeleton*.

THE BONES OF THE HEAD

For the purposes of description the skull (Fig. 10.2) may be divided into:

- The bones of the cranium
- The bones of the face.

The bones of the cranium

The cranium is a box-like cavity that contains and protects the brain. It has a dome-shaped roof called the *calvaria*, or skull cap, and its

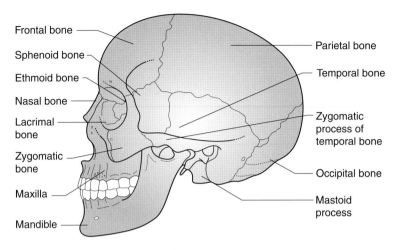

Fig. 10.2   The bones of the head.

floor is known as the base of the skull. The cranium consists of 15 bones:

- one frontal bone
- two parietal bones
- one occipital bone
- two temporal bones
- one ethmoid bone
- one sphenoid bone
- two inferior nasal conchae
- two lacrimal bones
- two nasal bones
- one vomer.

The *frontal bone* is a large flat bone comprising the forehead and most of the roof of the orbit. There are rounded prominences, called the *frontal tuberosities*, one on each side of the midline, which vary in size from one individual to another and which together form the forehead proper. The bone contains two irregular cavities called the *frontal sinuses*, which lie one over each orbit and which open into the nasal cavity. The sinuses contain air and are lined with mucous membrane, which is continuous with the mucous membrane lining the respiratory tract. They add resonance to the voice and they serve to lighten the skull, but the mucous membrane may become infected, causing a condition known as *sinusitis*.

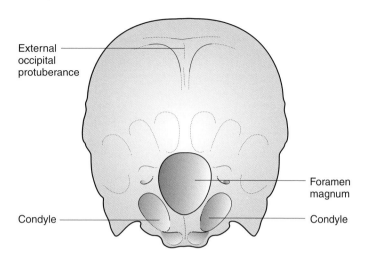

Fig. 10.3 The occipital bone (from below).

The *parietal bones* form the sides and roof of the cranium; they articulate with the frontal bone, the occipital bone and with each other to form the sutures or joints of the cranium (see Chapter 12). On the internal surfaces are small grooves to carry the blood vessels supplying the brain, and the impression of the folds or convolutions of the surface of the brain can be seen. At birth there are membranous gaps in the skull at the angles of the parietal bone which are called *fontanelles* (see Chapter 12).

The *occipital bone* (Fig. 10.3) forms the back of the skull. It carries a marked prominence, the external occipital protuberance, which provides attachment for muscles. Below this there is a large oval opening, the *foramen magnum*, through which the cranial cavity communicates with the vertebral canal. On either side of the foramen are two smooth oval processes, the *occipital condyles*, for articulation with the first cervical vertebra. This joint allows the nodding movement of the head.

The *temporal bones* (Fig. 10.4) are situated at the sides and base of the skull. Each consists of four parts:

1. The squamous part forms the anterior and upper part of the bone and is thin and flat. A long arched process, the zygoma or *zygomatic process*, projects forward from the lower portion.

2. The *petromastoid part* forms the posterior portion of the bone and can be divided further into two parts:
   a. The *mastoid portion*, which continues into a conical projection called the mastoid process, containing air cells.

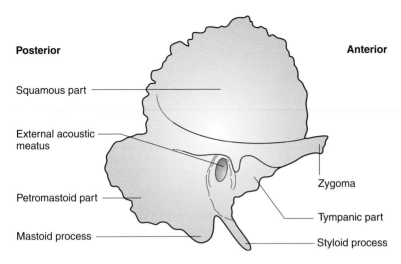

Posterior   **Anterior**

Squamous part

External acoustic
meatus

Petromastoid part

Mastoid process

Zygoma

Tympanic part

Styloid process

Fig. 10.4   The temporal bone.

b. The *petrous portion*, between the occipital bone and the sphenoid, which contains the structures forming the internal ear.

3. The *tympanic part* is a curved plate lying below the squamous
part and in front of the mastoid process. It contains the external
acoustic meatus.

4. The *styloid process* projects downwards and forwards from the
underside of the bone.

The *ethmoid bone* (Fig. 10.5) is very light and irregular in shape
and consists of three parts:

1. The *cribriform plate*, a small horizontal plate perforated by many
fine openings (foramina) for the passage of the olfactory
nerves, which transmit the sense of smell.

2. The *perpendicular plate*, which descends from the cribriform
plate and forms the upper part of the nasal septum, which
divides the nasal cavity into two.

3. Two *labyrinths*, each consisting of a number of thin-walled
ethmoidal air cells that communicate with the nasal cavity and
may become infected from it. Two thin plates of bone called the
superior and *middle nasal conchae* jut out into the nasal cavities
from the spongy labyrinths.

The *sphenoid bone* (Fig. 10.6) is situated at the base of the skull,
in front of the temporal bones. It is shaped rather like a bat with

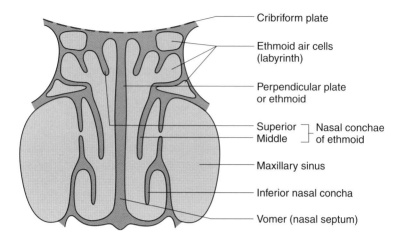

Fig. 10.5 Diagram of a section of the ethmoid bone (shown in colour).

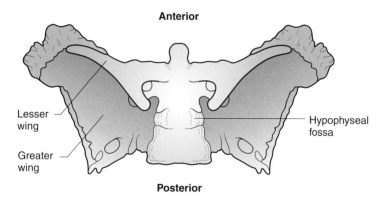

Fig. 10.6 The sphenoid bone.

outstretched wings. The body contains two large air sinuses, which communicate with the nasal cavity, and a deep depression, the *hypophyseal fossa*, which contains the *hypophysis cerebri* or pituitary gland. The greater and lesser wings are perforated by many openings for the passage of nerves and blood vessels.

The *inferior nasal conchae* are curved plates of bone that lie in the walls of the nasal cavity, below the superior and middle nasal conchae of the ethmoid bone (see Fig. 10.5).

The *lacrimal bones* are the smallest and most fragile of the cranial bones and form part of the walls of the orbits. Each is grooved to contain the lacrimal sac and nasolacrimal duct, by which the lacrimal fluid, or tears, which washes constantly over the surface of the eye, is carried into the nasal cavity.

The *nasal bones* are two small oblong bones that together form the bridge of the nose.

The *vomer* is a flat bone forming the lower part of the septum of the nose.

The bones of the face

The bones of the face are:

- The maxillae
- The mandible
- Two zygomatic bones
- Two palatine bones
- The hyoid bone.

The *maxillae* are the largest bones of the face (apart from the mandible) and by their union in the midline, form the whole of the upper jaw. They carry the upper teeth embedded in a ridge of bone called the alveolar process. The *palatine process* of the maxilla is a horizontal projection which forms a considerable part of the floor of the nasal cavity and the roof of the mouth. The maxillary sinus is an air-filled cavity within the body of the bone, which communicates with the nasal cavity and which may become infected following nasal infection.

The *mandible* (Fig. 10.7) is an irregular bone and is the only movable bone in the head. It forms the lower jaw and carries the lower teeth

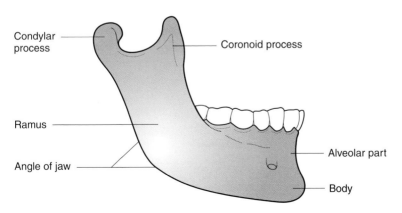

Condylar process

Coronoid process

Ramus

Angle of jaw

Alveolar part

Body

Fig. 10.7   The mandible.

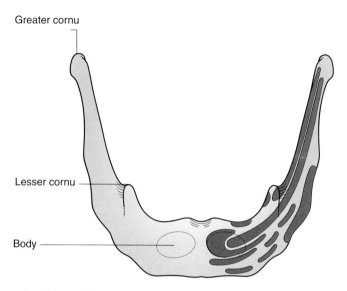

Greater cornu

Lesser cornu

Body

Fig. 10.8 The hyoid bone.

embedded in the alveolar part. The *rami* are vertical projections that carry a condylar process, which articulates with the temporal bone, and a coronoid process, which provides attachment for muscle. The upright and horizontal portions meet to form the angles of the jaw.

The *zygomatic bones* are irregular bones forming the prominence of the cheek and part of the walls of the orbit. The temporal process articulates with the zygomatic process of the temporal bone to form the zygomatic arch.

The *palatine bones* are irregular bones that form part of the hard palate, the lateral wall of the nasal cavity and the floor of the orbit.

The *hyoid bone* (Fig. 10.8) is U shaped and lies at the base of the tongue, to which it gives attachment. It does not articulate with any other bone but is connected by ligaments to the styloid processes of the temporal bone (see Fig. 10.4).

THE BONES OF THE TRUNK

The bones of the trunk are:

- The sternum
- The ribs
- The vertebral column.

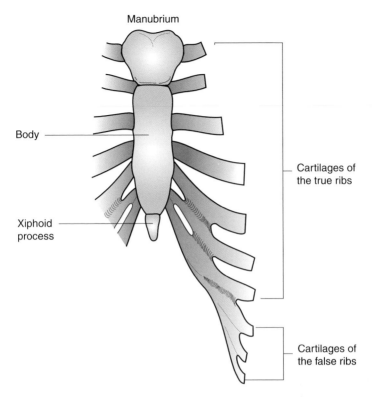

Fig. 10.9   The sternum and the costal cartilages.

The sternum

The sternum (Fig. 10.9) is a long flat bone that runs down the front of the thorax just under the skin. Its upper end supports the clavicles and it also articulates with the first seven pairs of ribs. This bone is in three parts:

1. The *manubrium* is roughly triangular and its lower border is covered with a thin layer of cartilage for articulation with the upper end of the body.

2. The body is longer and narrower than the manubrium. Where it joins the manubrium, there is a small notch, which accommodates the cartilage of the second rib.

3. The *xiphoid process* is small and variable in shape and may not become completely ossified.

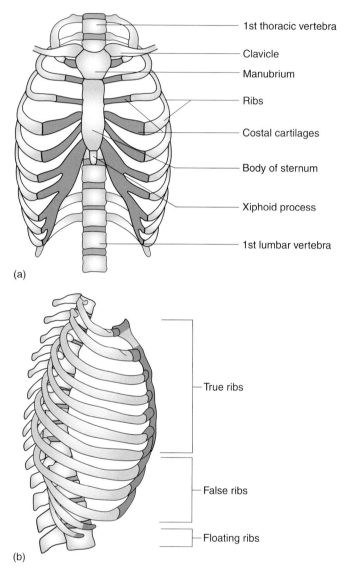

Fig. 10.10 The thoracic cage.

The ribs

The ribs are arched bones which are connected behind with the vertebral column. There are usually twelve pairs, the first seven of which are attached to the sternum by the *costal cartilages* and are known as

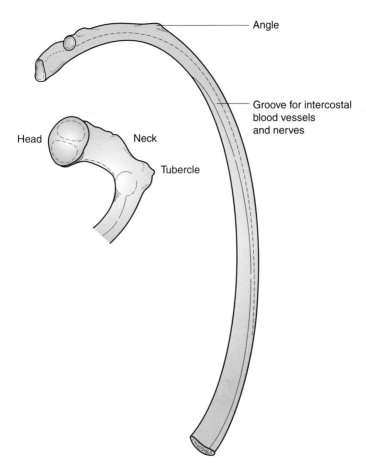

Fig. 10.11 A typical rib.

the true ribs. The remaining five are called false ribs. Of these the upper three are joined to the cartilage of the rib immediately above; the lower two are free at their anterior ends and are known as floating ribs.

The ribs form the curved walls of the *thorax*, sloping downwards towards the front. They increase in size from above downwards so that the thoracic cavity is roughly cone shaped (Fig. 10.10).

Each rib is curved and the undersurface is grooved for the passage of intercostal arteries, veins and nerves. The vertebral end has a head, a neck and a tubercle. The head has two smooth facets, which articulate with the bodies of the corresponding vertebrae (Fig. 10.11).

The *tubercle* has a small oval facet for articulation with the transverse process of the corresponding vertebra.

The vertebral column

The vertebral column consists of number of irregular bones called the *vertebrae*, which are firmly connected to one another but are capable of a limited amount of movement on one another. The column provides a central axis and also protects the spinal cord, which it surrounds. Each vertebra consists of a cylindrical body lying to the front, and a vertebral arch projecting backwards and enclosing a space called the *vertebral foramen* through which the spinal cord passes. The arch has a spinous process directed backwards and downwards and two transverse processes that project laterally. These are for the attachment of muscles and ligaments. On the undersurface of the arch is a notch for the passage of spinal nerves and vessels. Each vertebra has four *articular processes*, two above and two below, which meet the corresponding processes of adjoining vertebrae.

The wide parts of the arch carrying the spinous process are known as the *laminae*; this is the part removed in the operation of laminectomy to relieve pressure on the spinal cord caused by disease or following injury.

The bodies of adjoining vertebrae are firmly connected to one another by discs of fibrocartilage called *intervertebral discs*. Each disc has an outer ring of fibrocartilage and an inner core called the *nucleus pulposus*. It is possible for the outer ring to become weakened and for the nucleus pulposus to irritate an adjoining nerve root causing pain.

The vertebrae are divided into five groups:

- seven cervical vertebrae
- twelve thoracic vertebrae
- five lumbar vertebrae
- five sacral vertebrae
- four coccygeal vertebrae.

The seven *cervical vertebrae* are the smallest and can be easily identified because their transverse processes are perforated by foramina for the passage of the vertebral arteries (Fig. 10.12). The spinous process is forked and provides attachment for muscles and ligaments.

The first cervical vertebra is called the *atlas*. It has no body or spine but consists of a ring of bone with two facets, which articulate with the occipital bone (Fig. 10.13). A ligament called the transverse ligament divides the ring of the atlas into two.

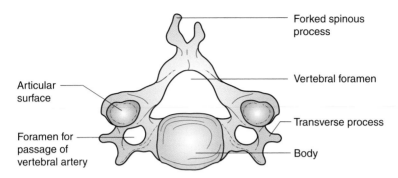

Fig. 10.12 A typical cervical vertebra.

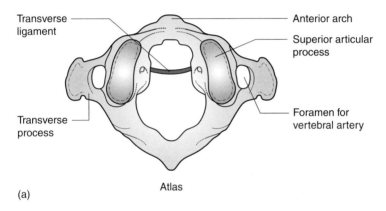

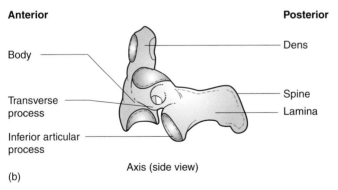

Fig. 10.13 The first and second cervical vertebrae.

The *axis*, or second cervical vertebra, carries a strong tooth-like process called the dens (or odontoid process), which projects upwards from the body and provides a pivot on which the atlas, and therefore the skull, rotates, allowing a turning movement of the

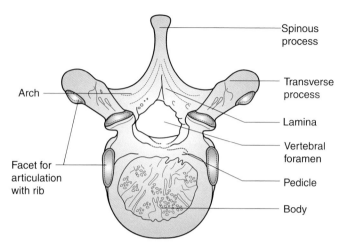

Fig. 10.14 A typical thoracic vertebra (from above).

head (see Fig. 10.13). The dens is retained in position by the transverse ligament of the atlas, behind which lies the spinal cord.

The seventh cervical vertebra is distinguished by a very long spinous process which can be seen and felt through the skin at the base of the neck and which is not forked.

The twelve *thoracic vertebrae* are larger than the cervical vertebrae and they increase in size from above downwards. The body is roughly heart shaped and has two distinguishing features:

1. Additional facets on each side to articulate with the head and tubercle of the corresponding rib (Fig. 10.14).
2. Long pointed spinous processes, which project downwards and backwards (Fig. 10.15).

The heads of the ribs lie between the vertebrae and articulate with one facet on the vertebra above and one on the vertebra below.

The five *lumbar vertebrae* are the largest vertebrae and have no facets for articulation with the ribs. The spinous processes are large and strong and provide attachment for the muscles (Fig. 10.16).

The five *sacral vertebrae* are fused together to form a large bone, the sacrum (Fig. 10.17), which is roughly triangular and forms a wedge between the two hip bones with which it articulates. The pelvic surface of the bone is concave and the anterior projection at the upper end is known as the sacral promontory (Fig. 10.18). The vertebral foramen found in the other vertebrae is here called the

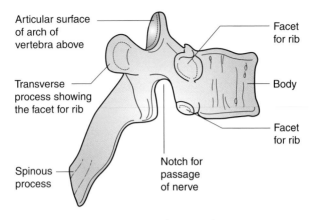

Articular surface
of arch of
vertebra above

Facet
for rib

Transverse
process showing
the facet for rib

Body

Facet
for rib

Spinous
process

Notch for
passage
of nerve

Fig. 10.15 A typical thoracic vertebra (side view).

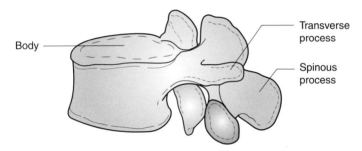

Body

Transverse
process

Spinous
process

Fig. 10.16 A lumbar vertebra.

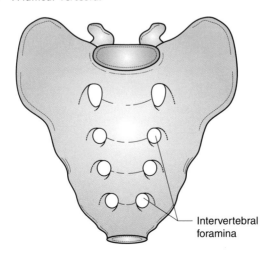

Intervertebral
foramina

Fig. 10.17 The sacrum.

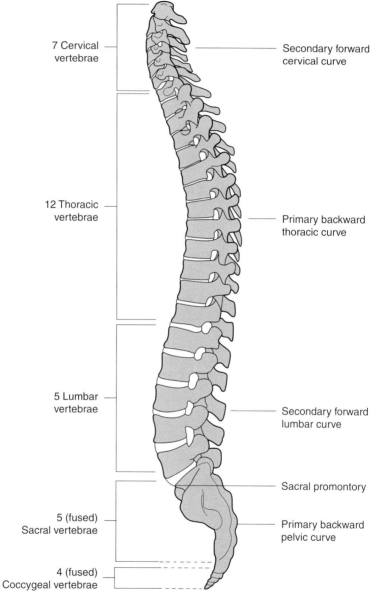

7 Cervical vertebrae

Secondary forward cervical curve

12 Thoracic vertebrae

Primary backward thoracic curve

5 Lumbar vertebrae

Secondary forward lumbar curve

Sacral promontory

5 (fused) Sacral vertebrae

Primary backward pelvic curve

4 (fused) Coccygeal vertebrae

Fig. 10.18 The curves of the spine.

sacral canal and, from it, four openings allow for the passage of nerve roots.

The *coccyx* is a small triangular bone consisting of four vertebrae fused together. It articulates with the sacrum and the joint allows slight movement backwards and forwards. In females this increases the size of the pelvic outlet during childbirth.

The vertebral column is the main support of the head and trunk as well as protecting the spinal cord. When viewed from the side it has four curves (see Fig. 10.18). The thoracic and pelvic curves are termed primary curves as they are present during fetal life. The cervical and lumbar curves are secondary as they appear or are accentuated when the child begins to hold up its head and sit up (cervical) and when it begins to stand and walk (lumbar).

There is only limited movement between any two adjoining vertebrae but there is considerable movement in the vertebral column as a whole. The intervertebral discs cushion any jarring that may occur as, for example, jumping from a height and landing on the feet. The curves of the spine enable it to bend without breaking although a blow on the column is more likely to cause a fracture or dislocation as the vertebrae are so firmly united to one another.

There are some abnormal curvatures of the spine including:

- Scoliosis, where the spine is twisted
- Kyphosis, where there is a hump at the top of the spine ('hunchback')
- Lordosis, where the spine is bent backwards giving the person an abnormally erect appearance.

SELF-TEST QUESTIONS

- What is the foramen magnum?
- What is the function of the sinuses?
- Describe the rib cage.
- What are the regions of the spine?
- Describe the curvatures of the spine.

Chapter 11

Bones of the limbs

LEARNING OBJECTIVES

After reading this chapter you should understand:

* the bones comprising the upper limbs
* the bones comprising the lower limbs

BONES OF THE UPPER LIMB

The bones of the upper limb are:

* The scapula ⎫
* The clavicle ⎭ forming the shoulder girdle
* The humerus
* The radius ⎫
* The ulna ⎭ forming the forearm
* Eight carpal bones
* Five metacarpal bones
* Fourteen phalanges.

The *scapula* (Fig. 11.1) is a triangular flat bone; it lies over the ribs at the back of the thorax but does not articulate with them. It is held in place by muscles which attach it to the ribs and vertebral column.

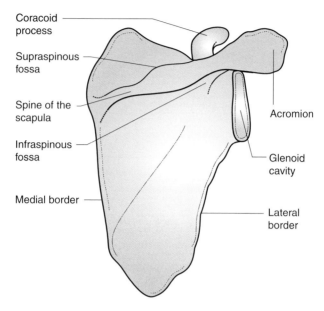

Fig. 11.1   The scapula.

This arrangement gives great freedom of movement to the shoulder girdle, making it possible to reach widely both forwards and backwards and to either side of the body. It is not often broken by falls as it is embedded in muscle. It has three borders and three angles; the lowest is spoken of as the angle because it is the sharpest and is easily felt. The front surface is concave to fit over the ribs; the posterior surface is convex and carries a projecting ridge of bone known as the spine of the scapula, which provides attachment for muscles and forms two depressions or fossae, one above and one below it.

The outer angle carries a shallow socket known as the *glenoid cavity*, which receives the head of the humerus to form the shoulder joint. Above this, two processes project:

1. The *acromion*, the larger of the two, which overlaps the socket and articulates with the clavicle to form the shoulder girdle.
2. The *coracoid process*, which juts forward and is like a hook.

Both of these can be easily felt. They provide attachment for muscles and help to keep the head of the *humerus* in place, preventing upward dislocation.

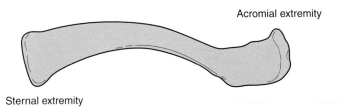

Acromial extremity

Sternal extremity

Fig. 11.2   The clavicle.

The *clavicle* (Fig. 11.2) is a long bone, roughly S shaped. It articulates with the sternum at its inner or sternal extremity and with the scapula at its outer or acromial extremity. The two extremities are easily distinguishable from one another: the inner is roughly like a pyramid in shape, while the outer is flatter and very similar in shape and form to the acromion process of the scapula, with which it articulates. The bone lies close under the skin and is easily felt along its whole course; starting from the sternal extremity, it curves first forwards and then backwards. It keeps the scapula in position and if it is broken the shoulder drops forwards and downwards. It is the only bony link between the bones of the upper limb and the axial skeleton, as the scapula does not articulate with either the ribs or vertebral column. It is a bone that is not found in the skeleton of many four-footed animals as it only becomes necessary to fix the scapula when the limb is moved outwards from the trunk. The bone is easily broken by falls on the shoulder as it is compressed between the sternum and the point of impact; it is, in fact, better that it should break than that there should be an injury at the root of the neck, where there are many important structures, or about the actual shoulder joint, where injury would be likely to limit subsequent movement.

The *humerus* (Fig. 11.3) is the largest and longest bone of the upper limb. The upper extremity has a hemispherical head, covered with hyaline cartilage, which articulates with the glenoid cavity of the scapula, forming the shoulder joint. The anatomical neck forms a slight constriction adjoining the head, and the greater and lesser tuberosities lie below the neck and provide attachment for muscles. Between the tuberosities, a deep groove accommodates one of the tendons of the biceps muscle. The shaft of the humerus has many roughened surfaces for the attachment of muscles, the most marked being the *deltoid tuberosity* on the outer side, which gives insertion to the deltoid muscle. A groove running obliquely round the shaft carries the radial nerve, one of the three main nerves of the upper limb.

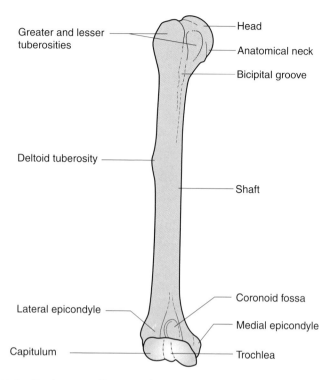

Greater and lesser tuberosities

Head

Anatomical neck

Bicipital groove

Deltoid tuberosity

Shaft

Lateral epicondyle

Coronoid fossa

Medial epicondyle

Capitulum

Trochlea

Fig. 11.3 The humerus (front view).

The lower end of the humerus is divided into articular and non-articular portions. The articular portion, together with the radius and the ulna, forms the elbow joint. It is divided by a shallow groove into the capitulum, a rounded projection that articulates with the radius, and the trochlea, which is shaped rather like a pulley and which articulates with the ulna. The non-articular portion has two epicondyles, which provide attachment for muscles. There are also two deep hollows: the posterior one is called the olecranon fossa as it accommodates the olecranon process of the ulna when the elbow is extended; and the coronoid fossa, which is on the anterior surface, provides for the coronoid process of the ulna when the elbow is flexed.

The *radius* (Fig. 11.4) is the outer bone of the forearm. The upper end is smaller and has a disc-shaped head with a hollowed upper surface to articulate with the capitulum of the humerus. The head also articulates with the ulna. The neck of the radius is the constricted portion below the head, and on the ulna side there is a projection

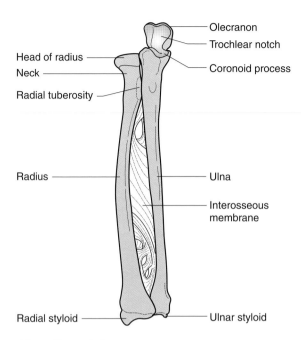

Fig. 11.4 The radius and ulna.

called the radial tuberosity, which gives insertion to the biceps muscle. The shaft of the bone has a sharp ridge facing the ulna and from it a sheet of fibrous tissue called the *interosseous membrane* runs to the ulna, connecting the two bones. The lower end of the radius is the wider part and comprises part of the wrist joint; it also has a projection called the *styloid process*, which can be felt at the base of the thumb.

The *ulna* (see Fig. 11.4) lies on the inner side of the forearm. The upper end is shaped like a hook and has two large projections; the *olecranon* fits into the olecranon fossa of the humerus when the arm is straight and its upper border forms the point of the elbow. It provides insertion for the tendon of the triceps muscle. The coronoid process is smaller and projects forwards. These two processes help in the formation of the trochlear notch, which articulates with the trochlea of the humerus. The radial notch is a depression on the upper part of the coronoid process, which articulates with the head of the radius and allows the turning movement of the hand. During rotation of the hand the lower end of the radius is carried round the lower end of the ulna so that the shafts of the two bones cross each other in the middle of the forearm.

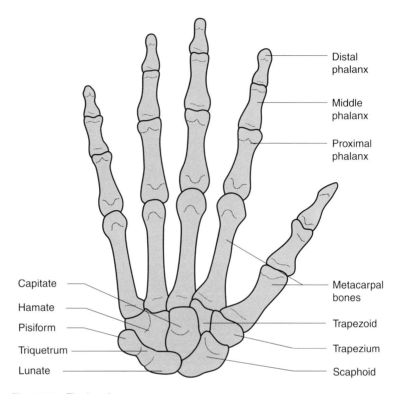

Distal phalanx

Middle phalanx

Proximal phalanx

Capitate

Hamate

Pisiform

Triquetrum

Lunate

Metacarpal bones

Trapezoid

Trapezium

Scaphoid

Fig. 11.5   The hand.

The shaft of the ulna, like that of the radius, carries a sharp ridge for the attachment of the interosseous membrane, which lies between the two bones.

The lower end of the ulna has a rounded part, known as the head, and a projection called the styloid process. The head articulates with the ulnar notch of the radius. The styloid process provides attachment for a ligament of the wrist joint; it may be felt below the skin of the wrist and is sometimes quite prominent.

The *carpal bones* (or carpus) (Fig. 11.5) comprise eight bones arranged in two rows of four. The bones of the upper row are named the *scaphoid*, the *lunate*, the *triquetrum* and the *pisiform*. The first three articulate with the radius. The bones of the lower row are called the *trapezium*, the *trapezoid*, the *capitate* and the *hamate*. The surface of the carpus, which forms the palm, has a deep cavity, the carpal groove, which has a fibrous band across it and which accommodates the median nerve and some of the tendons to the hand. This is known as the carpal tunnel.

The *metacarpal bones* (or metacarpus) are five miniature long bones running across the palm of the hand. The bases of the bones articulate with the lower bones of the carpus; the heads articulate with the phalanges. The first metacarpal, which articulates with the two phalanages which form the thumb, can be moved more freely than the other four and can be opposed to each finger in turn, thereby increasing the power of the grip.

The *phalanges* are also miniature long bones, three for each finger and two for the thumb.

BONES OF THE LOWER LIMB

The bones of the lower limb are:

- The hip bone, which forms part of the pelvis
- The femur
- The patella
- The tibia
- The fibula
- Seven tarsal bones
- Five metatarsal bones
- Fourteen phalanges.

The *hip bone* (Fig. 11.6) is a large, irregularly shaped bone, which articulates in front with the corresponding bone of the opposite side. Each bone consists of three parts – the *ilium*, the *ischium* and the *pubis* – which are united at the deep cavity on the outer aspect of the bone called the *acetabulum*. Full ossification is not completed until between the ages of 15 and 25 and before the bones are united by cartilage.

The *ilium* includes the upper part of the acetabulum and the expanded flattened area of bone above it, which is called the *iliac crest* and which provides attachment for the lateral muscles of the abdominal wall. The crest projects a little beyond the lower part of the bone and ends in front at the anterior *superior iliac spine*, which is easily felt at the lateral end of the fold of the groin. The posterior superior iliac spine underlies a small dimple, which is readily seen on the lower part of the back. The ilium also carries the anterior and posterior inferior iliac spines, which provide attachment for powerful muscles. At the back there is an articular surface for articulation

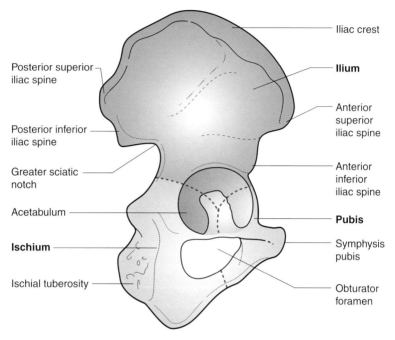

Fig. 11.6 The hip bone.

with the sacrum and below this there is a notch, the greater sciatic notch, for the passage of part of the sciatic nerve.

The *ischium* forms the lower posterior part of the hip bone. It carries the ischial tuberosity, which provides attachment for muscles and supports the body in the sitting position.

The *pubis* forms the anterior part of the hip bone and meets the pubis of the opposite side in a cartilaginous joint called the *symphysis pubis*. The pubis consists of a body, which enters the symphysis, and two branches, one running upwards to join the ilium and the other downwards to join the ischium. Between these branches and the ischium is a large opening, the *obturator foramen*, which is filled with a sheet of fibrous tissue.

The *acetabulum* is the deep socket in the centre of the lower part of the hip bone into which the head of the femur fits.

The *pelvis* is a bony ring composed of the two hip bones with the sacrum and coccyx behind. It is divided into the greater (false) and the lesser (true) pelvis by the *linea terminalis* and the *promontory* of the sacrum. The *greater pelvis* is the upper expanded portion bounded on each side by the ilium and at the back by the base of the sacrum. The *lesser pelvis* consists of a short curved canal deeper at

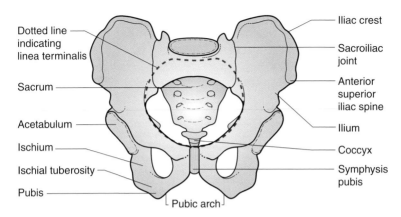

Dotted line indicating linea terminalis

Sacrum

Acetabulum

Ischium

Ischial tuberosity

Pubis

Iliac crest

Sacroiliac joint

Anterior superior iliac spine

Ilium

Coccyx

Symphysis pubis

Pubic arch

Fig. 11.7   The male pelvis.

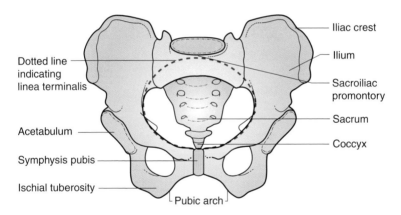

Dotted line indicating linea terminalis

Acetabulum

Symphysis pubis

Ischial tuberosity

Iliac crest

Ilium

Sacroiliac promontory

Sacrum

Coccyx

Pubic arch

Fig. 11.8   The female pelvis.

the back than the front. The female pelvis is shorter than that of the male and is also wider. The brim of the female pelvis is larger and more nearly circular than that of the male which is more typically heart shaped (Figs 11.7 and 11.8).

The *femur* (Fig. 11.9) is the longest and strongest bone in the human body. The upper end has a hemispherical head, which articulates with the acetabulum of the hip; near its centre is a small depression called the *fovea* (see Fig. 12.6), which provides attachment for the ligament of the head of the femur. This ligament runs to the base of the acetabulum. The neck of the femur projects at a marked angle from the shaft and enables free movement of the hip joint. Where the neck joins the shaft there are two processes, the *greater* and *lesser trochanters*,

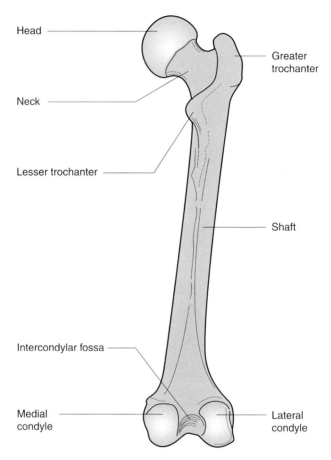

Head

Greater trochanter

Neck

Lesser trochanter

Shaft

Intercondylar fossa

Medial condyle

Lateral condyle

Fig. 11.9   The femur (back view).

which provide attachment for muscles. The greater trochanter is on the outer side and can be felt under the skin. The shaft of the femur is thinnest in the middle and widens considerably at the lower end. The posterior border is formed by a roughened ridge called the *linea aspera*, which provides attachment for muscles.

The lower end of the femur is widely expanded so that there is a good area for the transmission of body weight to the tibia. It has two prominent condyles, which articulate with the tibia. They are separated at the back by a deep gap called the *intercondylar fossa* and united in front by a smooth surface that articulates with the patella. Above the condyles at the back of the bone is the popliteal surface, which forms part of the *popliteal fossa*, containing blood vessels and nerves.

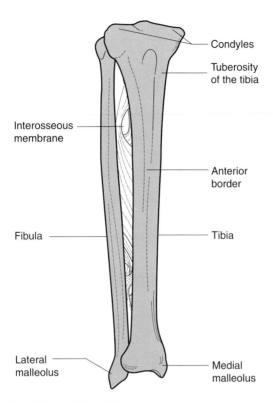

Fig. 11.10   The tibia and fibula (front view).

The *patella* is situated in front of the knee joint in the tendon of the quadriceps muscle, which straightens the knee. Bones that develop in tendons in this way are called sesamoid bones. The patella is flattened and triangular with the apex downwards. The posterior surface is smooth and articulates with the condyles of the femur. The anterior surface is roughened and is separated from the skin by a sac, similar to synovial membrane, called a *bursa*.

The *tibia* (Fig. 11.10) is the stronger of the two bones of the lower leg and lies on the inner (or medial) side. The upper end is greatly expanded to provide a good bearing surface for the body weight. It has two prominent masses, called the *medial* and *lateral condyles*, which are smooth and articulate with the condyles of the femur. Between the condyles is a roughened area that provides attachment for the ligaments and cartilages of the knee joint. Below the condyles is a smaller projection, called the *tuberosity of the tibia*, which provides attachment for the ligamentum patellae. The lateral condyle has a small circular facet for articulation with the upper end of the fibula.

The shaft of the tibia is roughly triangular in cross-section. The anterior border lies immediately under the skin and can be felt as the shin. A second border faces the fibula and provides attachment for the *interosseous membrane,* which connects the tibia and fibula, as the ulna and radius are connected in the forearm.

The lower end of the bone is slightly expanded and projects downwards to form the *medial malleolus,* on the inner side of the ankle, which articulates with the talus. The bone also articulates with the fibula.

The *fibula* (see Fig. 11.10) is very slender compared with the tibia and lies on the outer side of the leg. The head of the fibula has a circular facet that articulates with the *lateral condyle of the tibia* but does not comprise part of the knee joint. The shaft is slender and ridged and one of the ridges provides attachment for the interosseous membrane connecting the tibia and fibula. The lower end of the fibula projects downwards to a lower level than the tibia and carries the bony prominence on the outside of the ankle, known as the *lateral malleolus,* which articulates with the talus.

The *tarsal bones* (or tarsus) (Fig. 11.11) comprise seven bones, which make up the posterior half of the foot. The *talus* is the principal connecting link between the foot and the leg and forms an important part of the ankle joint. The *calcaneus* is the largest and strongest of the tarsal bones; it projects backwards to form the prominence of the heel and to provide a lever for the muscles of the calf, which are inserted into its posterior surface. The *navicular bone* lies between the talus and the three *cuneiform bones.* The cuneiform bones are wedge shaped and articulate with the navicular and the first three metatarsal bones. The cuboid bone is situated between the calcaneus and the fourth and fifth metatarsal bones.

The *metatarsal bones* (or metatarsus) are five miniature long bones, resembling the metacarpus. The bases articulate with the *cuneiform* and *cuboid bones.* The heads articulate with the *phalanges.* As in the hand the phalanges of the foot correspond in number and arrangement, there being only two in the big toe and three in each of the other toes.

The arches of the foot

The foot has two main functions: to support the weight of the body and to propel the body forward when walking. To fulfil these functions the foot has two *longitudinal arches*: the medial arch, which

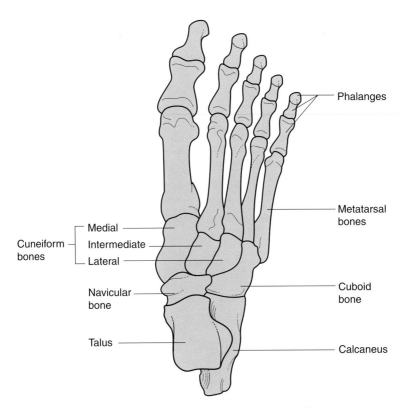

Fig. 11.11   The foot.

is particularly resilient; and the lateral arch, which is strong and allows only limited movement. There is also a series of transverse arches.

The arches of the foot give spring to the walk. They are helped in this by the strong ligaments, tendons and muscles that cross the sole of the foot and which, if stretched, may allow lowering of the medial longitudinal arch and eventually changes in the bones themselves, a condition known as 'flat foot'.

SELF-TEST QUESTIONS

- Describe the structure and movement of the forearm.
- What is the difference between the male and female pelvis?
- Describe the structure and movement of the lower limbs.
- What is an interosseous membrane and where is it found?

Chapter 12

Joints or articulations

LEARNING OBJECTIVES

After reading this chapter you should understand:

- the different groups of joints
- the function of joints
- the types of synovial joint

A joint or articulation is formed wherever two bones meet but not all junctions of bones allow movement. This chapter looks at the different types of joint in the human body and the types of movement that are possible at each. There are three groups of joints:

- Fibrous or fixed joints
- Cartilaginous or slightly movable joints
- Synovial or freely movable joints.

Not all joints fit rigidly into this classification as there are some fibrous joints that are slightly movable (e.g. the joint between the lower ends of the tibia and fibula) and some cartilaginous joints that are barely movable (e.g. the symphysis pubis).

Bones are joined to each other by ligaments, which are usually strong cords of fibrous tissue attached to the periosteum and running from one bone to another. These ligaments are yielding but inelastic and vary in strength and shape according to the work they have to do. Because they are pliable, they allow movement to take place but they are also strong, inelastic and rich in sensory nerves and, in this way, they protect the joints from excessive movement and strain.

FIBROUS JOINTS

One type of fibrous joint (Fig. 12.1) occurs where the margins of two bones meet and dovetail accurately into one another, separated only by a thin band of fibrous tissue. Joints such as these, which do not normally permit movement, are found between the bones of the cranium and are called *sutures*. In the infant at birth, there is a definite line of fibrous tissue between two adjoining bones, which allows the edges to glide over one another, enabling the head to be moulded to ease its passage through the birth canal. Other fibrous joints occur where the roots of the teeth articulate with the upper and lower jaws and where there is an interosseous ligament, as in the tibiofibular joint.

CARTILAGINOUS JOINTS

A cartilaginous joint (Fig. 12.2) occurs where the two bony surfaces are covered with hyaline cartilage and are connected by a pad of fibrocartilage and by ligaments, which do not form a complete capsule round the joint. A limited degree of movement is possible because the cartilaginous pad can be compressed. The joints between the bodies of the vertebrae and between the manubrium and the body of the sternum are cartilaginous joints.

Fibrous tissue

Bone

Fig. 12.1   Diagram of a fibrous joint.

SYNOVIAL JOINTS

A synovial joint (Fig. 12.3) consists of two or more bones, the ends of which are covered with articular hyaline cartilage. There is a joint cavity, containing *synovial fluid*, which nourishes the avascular articular cartilage, and the joint is completely surrounded by a fibrous capsule, lined with *synovial membrane*, which lines the whole of the interior of the joint with the exception of the bone ends, menisci and discs. The bones are also connected by a number of ligaments, and some movement is always possible in a synovial joint even though

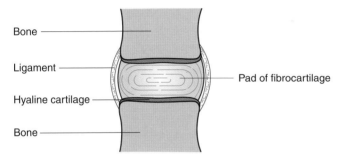

Bone

Ligament

Hyaline cartilage

Bone

Pad of fibrocartilage

Fig. 12.2   Diagrammatic section through a cartilaginous (slightly movable) joint.

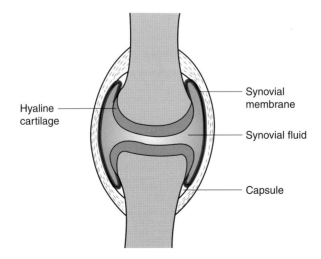

Hyaline cartilage

Synovial membrane

Synovial fluid

Capsule

Fig. 12.3   Diagrammatic section through a synovial (freely movable) joint.

it may be limited, as in the gliding movement between the adjoining metacarpal bones.

In some synovial joints the cavity may be divided by an articular disc or meniscus consisting of fibrocartilage, which helps to lubricate the joint and reduce wear of the articular surfaces.

Types of synovial joint

The synovial joints are divided into several classes according to the type of movement that occurs:

- Hinge joints – allow movement in one direction only; the elbow is a good example

- Pivot joints – allow rotation, e.g. the radius and ulna at the elbow allowing turning movements at the wrist, and the first and second cervical vertebrae allow turning movements of the head

- Condylar joints – two pairs of articular surfaces allow movement in one direction only but the articular surfaces may be enclosed in the same or in different capsules; the knee joint is one example

- Ball-and-socket joints – these are formed by a hemispherical head fitting into a cup-shaped socket; examples are the hip and shoulder joints

- Plane joints – gliding movements are restricted by ligaments or bony prominences, e.g. as in the carpal and tarsal joints.

JOINT MOVEMENTS

Joint movements are of three kinds: gliding; angular; and circular:

1. *Gliding* movement occurs, as its name implies, without any angular or rotatory movement.

2. *Angular* movement brings about an increase or decrease of the angles between the bones. It includes flexion or bending and extension or straightening, and also *abduction* (i.e. movement away from the midline) and *adduction* (i.e. movement towards the midline).

3. *Circular* movement allows internal rotation (i.e. turning a part on its own axis towards the midline) and external rotation (i.e. away from the midline). *Circumduction* allows movement of

a limb through a circle. The words *supination* and *pronation* refer to turning the hand palm uppermost or palm downwards, respectively.

These movements are often combined to produce a variety of movements at any one joint.

Different types of joint have varying kinds of movement. The ball-and-socket joints are most freely movable allowing all movements described above, excluding supination and pronation. Hinge joints allow flexion and extension only. In gliding joints, there is only slight movement increasing the range in all directions, as in the carpal and tarsal bones at the wrist and ankle, respectively. In the hand, the terms 'adduction' and 'abduction' are used to refer to movement to and from the midline of the part and not from the midline of the body as a whole. Adduction of the thumb brings it to and across the palm of the hand, and adduction of the little finger brings it towards the thumb. In the anatomical position, with the palm facing forward, the little finger is moving away from the midline of the body but towards the midline of the hand.

Joints are movable but the movements are carried out by the various muscles with which the joints are provided. The muscles, in addition to producing movement, run, as do the ligaments, from bone to bone and help to hold the bones in position and give support to the joint capsule, as long as their normal tone is sustained. If the muscles are paralysed and limp, the looser joints are dislocated comparatively easily, particularly the freely movable shoulder joint; if the muscles are paralysed, but rigid and shrunken, the joint may become completely immovable. This immobility can be prevented by moving the joint through as wide a range of movement as possible to maintain the elasticity of the muscles. Immobility may also result from the joint surfaces becoming adherent to one another or to the joint capsule as a result of disease or injury affecting the muscle itself.

THE JOINTS OF THE HEAD

The *temporomandibular joint*, between the temporal bone and the head of the mandible, is the only movable joint of the head and it is peculiar in that movement can occur in all three planes: upwards and downwards; backwards and forwards; and from side to side.

The sutures between the skull bones have already been described (see Chapter 10). However, at the angles of the parietal bones there are unossified membranous areas called fontanelles (Fig. 12.4).

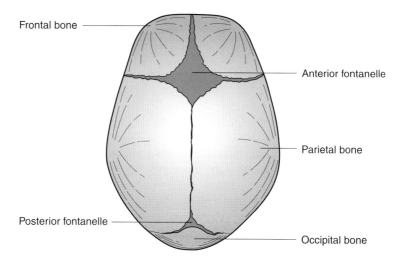

Fig. 12.4   An infant's head at birth, showing the fontanelles.

The *anterior fontanelle* is the largest and lies at the junction of the two parietal bones with the frontal bone. It is diamond shaped and does not close completely until the age of 15 to 18 months. Dehydration in infancy causes the fontanelle to become depressed, which is a serious sign. A large venous sinus runs under the fontanelle, from which it is possible to obtain a specimen of blood and into which intravenous fluids may be given to the infant.

The *posterior fontanelle* lies at the junction of the parietal bones with the occipital bone. It is triangular and closes shortly after birth. Delay in the closure of the fontanelles may be caused by *hydrocephalus* but this also sometimes occurs in the normal child.

THE JOINTS OF THE TRUNK

There are joints between all the vertebrae from the second cervical to the sacrum. *Cartilaginous joints* lie between the vertebral bodies, and synovial joints lie between the vertebral arches. Because the joints are so numerous the spinal column as a whole has considerable movement. Anterior and posterior longitudinal ligaments pass from the top of the spine to the sacrum, giving support. Other ligaments pass between the vertebral arches.

Between the ribs and the vertebrae are the *costovertebral joints*, which allow gliding movements; the same type of movement occurs at the sternocostal joints.

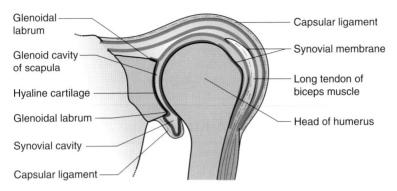

Glenoidal labrum

Glenoid cavity of scapula

Hyaline cartilage

Glenoidal labrum

Synovial cavity

Capsular ligament

Capsular ligament

Synovial membrane

Long tendon of biceps muscle

Head of humerus

Fig. 12.5   Diagrammatic section through the shoulder joint.

THE JOINTS OF THE UPPER EXTREMITY

The *sternoclavicular joint* is formed by the sternal end of the clavicle, the manubrium of the sternum and the cartilage of the first rib. It allows a gliding movement of the clavicle.

The *acromioclavicular joint* lies between the acromial extremity of the clavicle and the acromion of the scapula and is usually associated with movements of the shoulder.

The shoulder joint is a ball-and-socket joint (Fig. 12.5) and is the most freely movable of the joints in the body. It is formed by the head of the humerus fitting into the small, shallow glenoid cavity. The articular surfaces are covered with articular cartilage and the glenoid cavity is enlarged and deepened by a rim of fibrocartilage, called the *glenoidal labrum,* which runs around it. This lessens the risk of dislocation without limiting the movement as much as a larger and deeper bony socket would do. The bones are held together by a loose capsule of ligaments to give the joint its wide range of movement but the powerful muscles help to keep the bones in position. The long tendon of the biceps muscle serves as an intracapsular ligament. It runs through the bicipital groove between the tuberosities of the humerus into the joint cavity and, since it arises from the scapula immediately above the glenoid cavity, it tends to hold the articular surfaces in position. Movement of the arm above the level of the shoulder joint is due to movement of the scapula over the back of the thorax.

The elbow joint is complicated by the fact that it contains the hinge joint between the humerus and the radius and ulna and the pivot joint between the radius and ulna. There are strong ligaments running between the three bones and a circular ligament called the annular ligament, which keeps the head of the radius in contact with the radial notch of the ulna. The lower end of the radius also forms a pivot joint with the ulna.

The wrist joint is formed by the lower end of the radius and the scaphoid, lunate and triquetral bones. Together with the joints between the carpal bones the movements of flexion, extension, adduction (ulnar deviation), abduction (radial deviation) and circumduction can be carried out.

The *metacarpophalangeal joints* are also capable of all the movements described for the wrist joints but the interphalangeal joints are hinge joints, allowing flexion and extension only.

THE JOINTS OF THE LOWER EXTREMITY

The *sacroiliac joint* is a synovial joint allowing a small amount of rotary movement during flexion and extension of the trunk.

The *symphysis pubis* is a cartilaginous joint that moves very little. However, during pregnancy the pelvic joints and ligaments are relaxed to allow slightly greater movement.

The hip joint (Fig. 12.6) is a *ball-and-socket joint* formed by the head of the femur fitting into the cup-shaped acetabulum. The joint

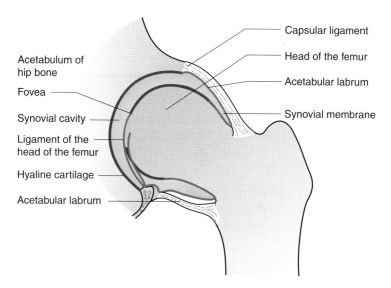

Fig. 12.6 Diagrammatic section through the hip joint.

surfaces are covered with articular cartilage and the acetabulum, like the glenoid cavity, is deepened by a rim of fibrocartilage called the acetabular labrum. The ligament of the head of the femur is attached to a small roughened pit called the fovea near the centre of the head of the femur, and runs to the acetabulum. The joint has a strong fibrous capsule and many ligaments, one of which, the iliofemoral ligament, lies across the front of the joint and prevents extension of the hip beyond a straight line with the trunk. A wide range of movement is possible at the hip joint, though when the knee is flexed, flexion of the hip is limited by the contact of the thigh with the anterior abdominal wall; also, when the knee is extended, flexion of the hip is limited by tension of the hamstring muscles.

The *knee joint* (Fig. 12.7) is the largest joint of the body. It is a compound joint: a condylar joint, which gives articulation between the condyles of the femur and the tibia, and a plane joint, which gives articulation between the patella and the femur. The joint has a fibrous capsule into the front of which the patella enters and which is lined with synovial membrane. The *cruciate ligaments* are very strong and cross each other within the joint. They run from the intercondylar area of the tibia to the femur and are partially covered with synovial membrane. The extracapsular ligaments are also

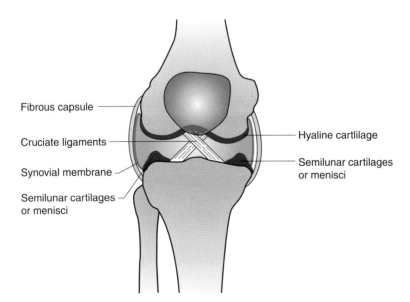

Fibrous capsule

Cruciate ligaments

Synovial membrane

Semilunar cartilages or menisci

Hyaline cartlilage

Semilunar cartilages or menisci

Fig. 12.7 Diagrammatic section through the knee joint.

strong and thick and help in controlling movement of the joint. The *menisci* (or semilunar cartilages) deepen the surfaces of the upper end of the tibia. They are wedge shaped, the outer border being thick and convex and the inner thin and concave, and they can be injured as a result of a twisting strain when the knee is flexed. However, if fully exercised, they will reform. The movements of the knee joint are mainly flexion and extension although some rotation can also occur.

The upper *tibiofibular joint* is a synovial plane joint allowing a little gliding movement, while at the lower end of the bones there is a small amount of rotation of the fibula during some ankle movements.

The *ankle joint* is a hinge joint formed between the tibia, fibula and talus (Fig. 12.8). The movements are flexion and extension but are usually spoken of as *dorsiflexion* (raising the foot) and *plantar flexion* (raising the heel).

The joints between the tarsal bones and between the tarsus and metatarsus are gliding joints and movement is limited. The *metatarsophalangeal joints* and the *interphalangeal joints* allow movements similar to the corresponding joints in the hand.

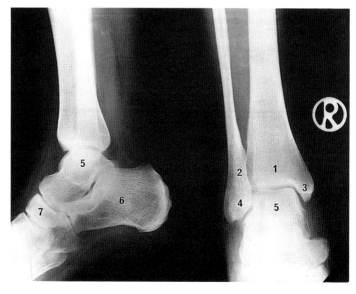

Fig. 12.8 X-rays of the right ankle joint, showing (left) lateral view and (right) anteroposterior view. (1) Tibia; (2) fibula; (3) medial malleolus; (4) lateral malleolus; (5) talus; (6) calcaneum; (7) cuboid bone.

Arthritis

There are two type of arthritis, both of which are painful and restrict movement at joints. *Osteoarthritis* results from degeneration of the articular surfaces of bone and mainly affects weight-bearing joints such as the hips. *Rheumatoid arthritis* affects joints such as those in the fingers and toes and arises from an inflammatory condition of the synovial membranes.

SELF-TEST QUESTIONS

- Which type of joint permits the greatest degree of movement?
- Compare and contrast a hinge and a ball-and-socket joint, giving examples.
- Describe the temporomandibular joint.
- What are the fontanelles?

Chapter 13

Muscle

CHAPTER CONTENTS

LEARNING OBJECTIVES

After reading this chapter you should understand:

- the structure of muscle
- the function of muscle
- the structure and function of the neuromuscular junction
- the muscles of the head
- the muscles of the neck
- the muscles of the trunk
- the muscles of the upper limb
- the muscles of the lower limb

STRUCTURE OF MUSCLE

The muscular system consists of a large number of muscles through which the movements of the body are carried out. Voluntary muscles are attached to bones, cartilages, ligaments, skin or to other muscles by fibrous structures called tendons and aponeuroses. The individual fibres of voluntary muscle with their sheaths of *sarcolemma* are bound together into bundles by the *endomysium* and are covered by the *perimysium*. The bundles, or *fasciculae*, are bound together by a denser covering called the *epimysium* and these groups form the individual voluntary muscles of the body (see Fig. 4.8). All muscles have a good blood supply from nearby arteries. Arterioles in the perimysium give off capillaries which run in the endomysium and across the fibres. Blood vessels and nerves enter the muscle together at the *hilum*.

Most muscles have *tendons* at one or both ends. Tendons are made of fibrous tissue and are usually cord-like in appearance, though in some flat or sheet-like muscles the cord is replaced by a thin, strong fibrous sheet called an *aponeurosis*. Fibrous tissue also forms a protective covering or muscle sheath, known as *fascia*.

Where one muscle is attached to another the fibres may interlace, the perimysium of one fusing with the perimysium of the other, or the two muscles may share a common tendon. A third type of connection occurs in the muscles of the abdominal wall where the fibres of the aponeuroses interlace, forming the *linea alba*, which can be seen as a shallow groove above the umbilicus.

ACTION OF MUSCLE

When a muscle contracts, one end normally remains stationary while the other end is drawn towards it (Figs 13.1 and 13.2). The end that remains stationary is called the origin and that which moves is called the insertion. It is not uncommon, however, for a muscle to be used, as it were, the wrong way round so that the insertion remains fixed and the origin moves towards it. The *gluteus maximus* provides an illustration. Its origin is in the sacrum and it is inserted into the femur. When the insertion moves towards the origin the flexed thigh is extended; when the body is bent forward at the hips the standing position is regained by movement of the origin towards the insertion. This arrangement economizes on the number of muscles required and further economy is achieved by the placing of muscles so that they can carry out more than one action. Muscles must cross

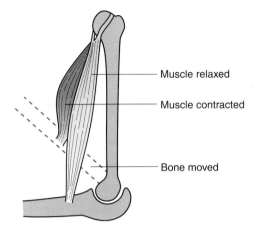

Fig. 13.1 Diagram showing bone movement as muscle contracts.

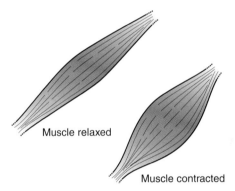

Fig. 13.2 A typical voluntary muscle showing contraction.

the joint they move; some cross two joints producing movement in both, e.g. the biceps crosses both elbow and shoulder causing flexion of both.

Muscles only act by contracting and pulling; they cannot push, though they can contract without shortening and so hold a joint firm and fixed in a certain position. When the contraction passes off, the muscles become soft but do not lengthen until stretched by the contraction of opposing muscles, known as *antagonists*.

Muscles never work alone – even the simplest movement requires the action of many muscles. Picking up a pencil requires the movement of fingers and thumb, wrist and elbow and possibly of the shoulder and trunk as the body leans forward. Each muscle must contract just sufficiently and each antagonist relax equally to allow the movement to take place smoothly without jerking. This concerted action of many muscles is termed *muscle co-ordination*. Any new action involving co-ordination requires time and practice until the new combination of muscle movement has been acquired and only then can it be carried out without great mental effort and concentration.

The sensory nerve gives 'muscle sense', which is not a very acute sensation but is sufficient to allow awareness of contraction and relaxation in the muscle. There is no awareness of this sensation until a conscious effort is made to relax or contract a muscle, at which time the previous degree of concentration becomes obvious. Under normal circumstances the muscles are in a state of partial contraction known as *muscle tone*; it is because of muscle tone that a position can be maintained for long periods without exhaustion. This is dependent on a mechanism whereby different groups of muscle fibres contract and relax in turn, giving periods of rest and activity to each group. The muscles having the highest degree of tonicity in humans are those of the neck and back.

Contraction of muscle

The composition of muscle is as follows:

- 75% water
- 20% protein
- 5% mineral salts, glycogen and fat.

Muscle contraction occurs as a result of *nerve impulses*. The nerve impulses, which are electrical, are transmitted to the muscle cells by chemical means and this is accomplished by the *neuromuscular junction* (Fig. 13.3). Nerve impulses arrive at the neuromuscular junction, which contains small packages of *acetylcholine*. This is released into the space between the nerve and the muscle, the *synaptic cleft*. When the acetylcholine attaches to the muscle cell it causes *depolarization* (see Chapter 5), and hence electrical activity spreads over the muscle cell leading to contraction.

In the disease myasthenia gravis, individuals release normal amounts of acetylcholine but the acetylcholine cannot attach to the muscle cells because of changes in the area of the muscle cell adjacent to the synaptic cleft. Individuals with myasthenia gravis have skeletal muscle weakness.

Energy is required for muscle fibres to contract; this is obtained from the oxidation of food, particularly carbohydrates. During

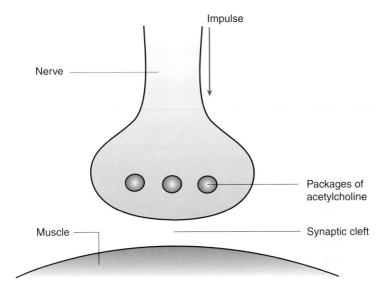

Fig. 13.3   The neuromuscular junction.

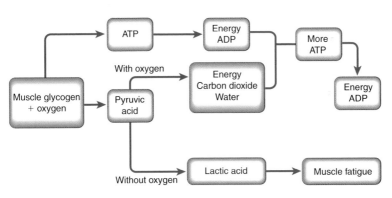

Fig. 13.4   Changes during muscle contraction.

digestion, carbohydrates are broken down to a simple sugar called glucose. The glucose that is not required immediately by the body is converted to glycogen and is stored in the liver and muscles. Muscle glycogen constitutes the source of heat and energy for muscular activity (Fig. 13.4). During the oxidation of glycogen to carbon dioxide and water, a compound is formed which is rich in energy. This compound is called *adenosine triphosphate* (ATP). When it is necessary for muscular contraction, the energy from ATP can be released as it changes to adenosine diphosphate (ADP). During the oxidation of glycogen, pyruvic acid is formed. If oxygen is plentiful, as it usually is during ordinary movement, pyruvic acid is broken down to carbon dioxide and water and, during the process, energy is released, which is used to make more ATP. If insufficient oxygen is available, the pyruvic acid is converted to *lactic acid*, which accumulates and produces *muscle fatigue*.

Skeletal muscle is also known as striated (striped) muscle (see Chapter 4) as a result of its microscopic appearance. The striations are due to the structure of the proteins of which the muscle is composed. These proteins are called *actin* and *myosin*. When muscle is contracted the striations are narrower – this is thought to result from the movement of one protein relative to the other in what is described as the *sliding filament theory* (Fig. 13.5). One of the proteins (myosin) has projections which, with the expenditure of energy, allow it to 'walk' along the other protein, thus leading to contraction when the muscle is stimulated by an electrical impulse.

During violent exercise more oxygen is brought to the muscles but, even so, not enough oxygen reaches the muscle cells, particularly at

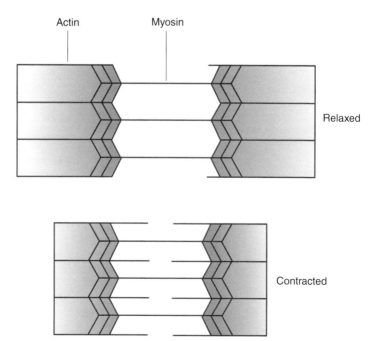

Fig. 13.5   The sliding filament theory.

the beginning of the effort. Lactic acid accumulates and diffuses into the tissue fluid and blood. The presence of lactic acid in the blood stimulates the respiratory centre, and the rate and depth of respiration are increased. This continues, even after the exercise is over, until sufficient oxygen has been taken in to allow the cells of the muscles and the liver to oxidize the lactic acid completely, or convert it to glycogen. The extra oxygen needed to remove the accumulated lactic acid is called the 'oxygen debt', which must be repaid after the exercise is completed.

CHIEF MUSCLES OF THE BODY

MUSCLES OF THE HEAD

The muscles of the head (Fig. 13.6) are divided into two groups according to their function: the muscles of expression and the muscles of mastication.

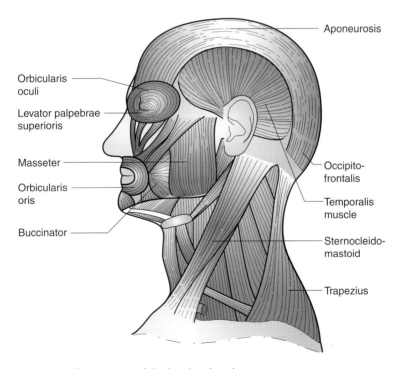

Fig. 13.6   The muscles of the head and neck.

The *muscles of expression* are attached to the skin rather than the bone so that they move the skin and change the facial appearance. Circular muscles, called *orbicularis oculi* and *orbicularis oris*, surround the eyes and mouth, respectively, closing them. Small muscles raise and lower the eyebrows and upper lids, raise and lower the angles of the mouth and dilate the nostrils, causing a look of surprise, worry, happiness or sorrow. Small muscles also move the eyeballs in the orbits (see Chapter 7) both to direct the eyes for sight and also to change the expression.

The *muscles of mastication* move the lower jaw up and down in biting, and also from side to side and backwards and forwards in chewing. These muscles are the *masseter* (running to the angle of the jaw from the zygomatic arch), the *temporalis muscle* (lying over the temporal bone and inserted into the lower jaw) and the smaller muscles, which also run from the skull to the lower jaw. These are the muscles that 'lock' the jaw in *tetanus*.

MUSCLES OF THE NECK

The neck contains two large muscles: the sternocleidomastoid and the trapezius (see Fig. 13.6).

The *sternocleidomastoid* lies at the side of the neck, running from the sternum and clavicle to the mastoid process and the surface of the temporal bone behind it. When the muscle on one side contracts, it draws the head towards the shoulder and turns the head. When both are used together, they flex the neck.

The *trapezius* lies over the back of the neck and shoulder and is roughly triangular, with the base joining the spine down the back of the neck and chest, from the occiput, to which it is also attached, downwards. The angle is inserted into the scapula and clavicle, over the top and back of the shoulder. It draws the shoulders back when used as a whole, and also draws the scapula up and down, when the upper and lower portions are used separately (Table 13.1).

MUSCLES OF THE TRUNK

The chief muscles of the trunk can be grouped according to their function:

- Muscles moving the shoulder
- Muscles of respiration
- Muscles forming the abdominal wall
- Muscles moving the hip

Table 13.1 Muscles of the neck

Name	Position	Origin	Insertion	Action
Sternocleidomastoid	The side of the neck	Sternum and clavicle	Mastoid process	Used separately, turn the head to the side and tilt to same shoulder; used together, flex the neck
Trapezius	The back of the neck and shoulder	The occiput and the spines of the thoracic vertebrae	The clavicle and the spine of the scapula	Draws the scapula back, bracing the shoulders; raising and lowering the shoulder and extending the neck by exerting pull on the occiput

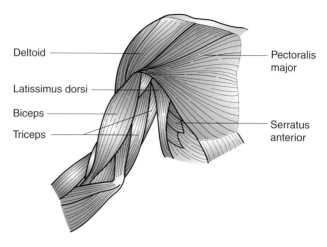

Deltoid

Latissimus dorsi

Biceps

Triceps

Pectoralis major

Serratus anterior

Fig. 13.7 The muscles of the shoulder and arm.

- Muscles moving the spine
- Muscles of the pelvic floor.

Muscles moving the shoulder

The chief muscles moving the shoulder are the powerful muscles covering the back and front of the chest. They include the *pectoralis major*, the *trapezius* (see p. 185), the *latissimus dorsi* and the *serratus anterior* (Fig. 13.7). The pectoralis covers the front of the chest, running from the sternum out to the humerus. The latissimus dorsi covers the back of the chest and abdomen, running from the thoracic and lumbar vertebrae and iliac crest out to the humerus (Fig. 13.8). These muscles form the muscle in front of and behind the armpit. The serratus anterior runs round the side wall of the thorax from the ribs in front to the vertebral edges of the scapula under which it passes (Table 13.2).

Muscles of respiration

The chief muscles of respiration are:

- The diaphragm
- The external intercostals
- The internal intercostals.

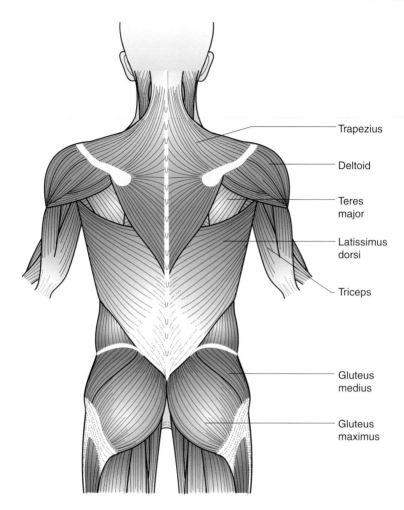

Fig. 13.8   The muscles of the back.

The *diaphragm* is a domed sheet of muscle dividing the thorax from the abdomen. The border is of muscle, while the centre is a sheet of fibrous tissue or aponeurosis. The muscle arises from the tip of the sternum, the lower ribs and their cartilages, and the first three lumbar vertebrae, and is inserted into the central aponeurosis. Three openings in the diaphragm allow for the passage of the oesophagus, aorta and inferior vena cava, together with lesser structures such as the vagus nerve and thoracic duct, which accompany the oesophagus

Table 13.2 Muscles connecting the upper limb with the trunk

Name	Position	Origin	Insertion	Action
Pectoralis major	Front of chest	Sternum, clavicle and cartilages of true ribs	Humerus (bicipital groove)	Adducts the shoulder, draws the arm across the front of the chest. Internal rotation of shoulder
Latissimus dorsi	Crosses the back from the lumbar region to the shoulder	Lower thoracic, lumbar and sacral vertebrae and iliac crest	Humerus (bicipital groove)	Adducts the shoulder, draws the arm backwards and downwards, as in bell-pulling and rowing, and internal rotation of shoulder
Serratus anterior	Over the sides of the thorax and under the scapula at the back	The front of the eight upper ribs	The medial border of the scapula	Draws the scapula forwards; antagonistic to the trapezius

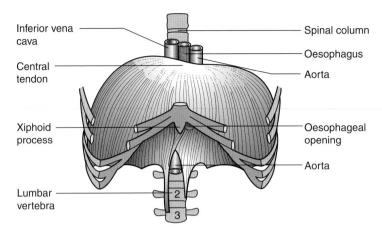

Inferior vena cava

Central tendon

Xiphoid process

Lumbar vertebra

Spinal column

Oesophagus

Aorta

Oesophageal opening

Aorta

Fig. 13.9   The diaphragm.

and aorta, respectively. When the muscle fibres contract, the dome of the diaphragm is flattened and lowered, which increases the depth of the thorax from top to bottom (Fig. 13.9).

The *external intercostal muscles* lie between the ribs, their fibres running downwards and forwards from one rib to the rib below. Each originates from the lower border of the rib above and is inserted into the upper border of the rib below. Their action is to draw the ribs upwards and outwards to increase the size of the thorax from side to side and from back to front.

The *internal intercostal muscles* also lie between the ribs under the external intercostals and are antagonistic to them. Their fibres run downwards and backwards from one rib to the rib below. Each originates from the lower border of the rib above and is inserted into the upper border of the rib below. Their action is to draw the ribs downwards and inwards to decrease the size of the thorax from side to side and from back to front, particularly in forced expiration.

Under conditions where breathing is made difficult, for example, in lung diseases such as *asthma*, muscles at the top of the trunk, other than those above, are used to increase the volume of the thorax. These are called the accessory muscles of respiration – the trapezius is an example of one of these muscles. Their use can be observed in people who have difficulty breathing because the shoulders rise and fall with inspiration and expiration.

Muscles forming the abdominal wall

The chief muscles of the abdominal wall (Fig. 13.10) are:

- The rectus abdominis, forming the front wall
- The external oblique
- The internal oblique } forming the side wall and lying
- The transversus abdominis } one under the other
- The quadratus lumborum.

The *rectus abdominis* forms the anterior abdominal wall, running up from the pubis to the sternum and costal cartilages. Its fibres run straight up and down, hence its name ('rectus' means straight). It is divided into two parts by a line of fibrous tissue, the linea alba, in the midline of the body. It is also crossed by lines of fibrous tissue at intervals. These fibrous bands strengthen the muscle and help to prevent stretching.

The *external oblique muscle* forms the outer layer of the side wall. Its fibres run downwards and forwards. It arises from the lower ribs and is inserted into the iliac crest and the inguinal ligament.

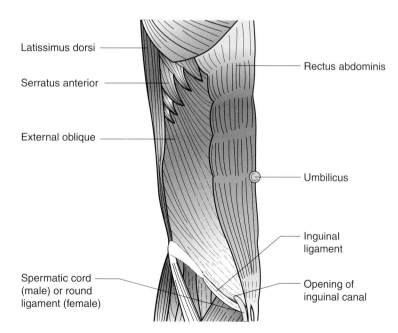

Latissimus dorsi

Serratus anterior

External oblique

Rectus abdominis

Umbilicus

Inguinal ligament

Spermatic cord (male) or round ligament (female)

Opening of inguinal canal

Fig. 13.10 The abdominal wall. Note that the aponeurosis has been cut away to expose the rectus.

The *inguinal ligament*, which is a strong cord of fibrous tissue, forms the firm edge of the abdominal wall across the groin, where the muscles are not inserted into the bone, leaving a gap for muscles, blood vessels and nerves to pass under it into the limb from the trunk. It provides attachment for the muscles here. Across the front of the abdomen the external oblique muscle forms a strong aponeurosis, which passes in front of the rectus, joining the linea alba.

The *internal oblique* forms the second layer of the side wall of the abdomen. Its fibres run upwards and forwards. It arises from the iliac crest and the inguinal ligament and is inserted into the lower ribs and their cartilages. It also forms an aponeurosis, passing partly in front and partly behind the rectus, and joining those of the external oblique and transversus abdominis.

The *transversus abdominis* forms the inner layer of the side wall of the abdomen, lying under the internal oblique. Its fibres run straight round the abdominal wall. It arises from the iliac crest and the lumbar fascia, by which it is joined to the lumbar vertebrae. It is inserted into an aponeurosis that runs across the front of the abdomen behind the rectus and joins the linea alba.

The *quadratus lumborum* forms the posterior wall, running up from the iliac crest to the twelfth rib and upper lumbar vertebrae. It holds the twelfth rib steady during breathing.

The abdominal wall is pierced by a canal in either groin – this is called the *inguinal canal*. It runs obliquely through the muscular coats just above the inguinal ligament near its inner extremity. The canal gives passage to structures: in the male, the spermatic cord from the testicle; in the female, the round ligaments of the uterus run through it with associated blood vessels and nerves.

Muscles of the hip

The muscles in the trunk moving the hip are:

- The iliopsoas
  - the psoas
  - the iliacus
- The gluteal muscles: maximus; medius; and minimus.

The *iliopsoas muscles* cross the front of the groin behind the inguinal ligament (Fig. 13.11). The *psoas* arises from the transverse processes of the lumbar vertebrae, and the *iliacus* from the front surface of the upper part of the ilium. They are both inserted into the lesser

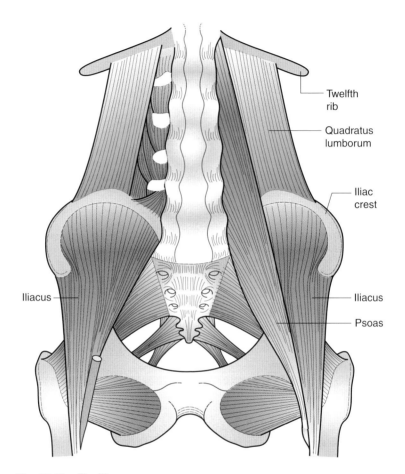

Fig. 13.11 The iliopsoas.

trochanter of the femur. They flex the hip joint and also produce abduction but when the femur is fixed, they bend the trunk forwards.

The *gluteal muscles* form the buttocks (see Fig. 13.8), running from the back of the sacrum and ilium to their insertion in the greater trochanter of the femur and the gluteal ridge below it. They are three in number: the *gluteus maximus, medius* and *minimus*. They extend the hip joint and also abduct and laterally rotate the hip (Table 13.3) but when the femur is fixed they extend the trunk on the lower limb. These muscles are commonly used as the site for *intramuscular injections* as they are thick and fleshy. (NB Care must be taken to use the *upper outer quadrant* as the sciatic nerve passes through the other quadrants.)

Table 13.3   Trunk muscles that move the hip

Name	Position	Origin	Insertion	Action
Psoas major	Crosses the groin behind the inguinal ligament	Transverse processes of lumbar vertebrae	Lesser trochanter of femur	Flexes the hip
Iliacus	Crosses the groin behind the inguinal ligament with the psoas	Front surface of the iliac bone	Lesser trochanter of femur	Flexes the hip
Gluteal muscles	Cross the back of the hip, forming the buttocks	Posterior surface of ilium and sacrum	Greater trochanter and gluteal line of femur	Extend, abduct and laterally rotate the hip

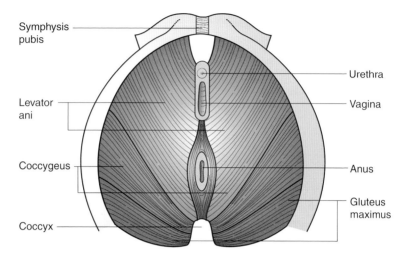

Symphysis pubis

Levator ani

Coccygeus

Coccyx

Urethra

Vagina

Anus

Gluteus maximus

Fig. 13.12 The pelvic diaphragm.

Muscles moving the spine

The muscles of the abdominal wall flex and turn the trunk, the rectus flexing and the side muscles turning the thorax on the abdomen. The abdominal muscles also compress the internal organs. The *spinalis muscles* extend the spine. They run up the back of the trunk on either side of the spine. They arise from the back of the iliac crest and the sacrum, and are inserted into the ribs and upper vertebrae.

Muscles of the pelvic diaphragm

The *pelvic diaphragm* consists of muscles that form the support of the pelvic organs (Fig. 13.12). It runs from the pubis in front, back to the sacrum and coccyx, and out to the ischium on either side. It is like an open book in shape, sloping downwards from back to front and from either side towards the midline. The pelvic diaphragm is composed of the *levator ani* and *coccygeus muscles*. It is pierced in the midline in the female by three openings for the passage of the urethra, vagina and rectum. In the male there are only two openings (for the urethra and rectum).

MUSCLES OF THE UPPER LIMB

The muscles of the upper limb may be divided into the muscles of the arm, forearm and hand.

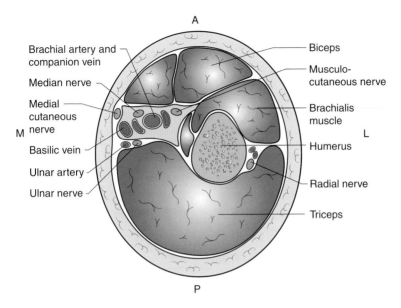

Fig. 13.13 Section of the arm, showing the relation of the muscles to the bone and other structures. L, Lateral; M, medial; A, anterior; P, posterior.

Muscles of the arm

The muscles of the arm, shown in Fig. 13.13, are the largest and strongest of the limb and include:

- The biceps
- The triceps
- The deltoid
- The brachialis.

The *biceps* is so called because it has two heads (from the Latin *caput* for head). It runs down the front of the arm, where it can be easily felt when it is contracted. It arises by two heads, one from the glenoid cavity and one from the coracoid process of the scapula, and is inserted into the radial tuberosity in the forearm, crossing the front of the elbow joint. It flexes the elbow and shoulder and supinates the hand. Hence the act of supination can be carried out with great force: screws and nuts are made so that a right-handed person tightens them by supination.

The *triceps* (so called because it has three heads) lies at the back of the arm. It arises by its three heads, one from the scapula and two from the humerus, and is inserted into the olecranon of the ulna, at the back of the elbow joint. It extends both elbow and shoulder and is antagonistic to the biceps.

The *deltoid* is approximately triangular and lies over the shoulder in the position of an epaulette with the base of the triangle forming the origin and attached to the shoulder girdle, just above the shoulder joint. It is inserted into the deltoid tuberosity on the outer side of the humerus. Its action is to abduct the shoulder to a right angle. (To raise the arm above a right angle, the shoulder girdle must move too, and the trapezius comes into play, drawing the scapula and clavicle up towards the occiput.) The front part of the deltoid muscle, used alone, helps to flex the shoulder, moving the humerus forwards, and the back portion, used alone, helps to extend the shoulder, moving the humerus backwards and towards the midline.

The *brachialis* lies lower in the front of the arm than the biceps, arising from the humerus and being inserted into the coronoid process of the ulna. It assists the biceps in the powerful action of flexion of the elbow (Table 13.4).

Muscles of the forearm

The forearm contains numerous small, less powerful muscles for movement of the wrist and digits (Fig. 13.14). In the front lie the *flexors of the wrist*, the *common flexors of the fingers*, the *long flexor of the thumb* and the *pronator muscles of the wrist*. The common flexors of the digits divide into four tendons, which run across the palm of the hand and up to the terminal phalanx of each digit, into which they are inserted. At the back lie the extensors of the wrist, the common extensors of the fingers, the extensor of the thumb and first finger, and the supinators of the wrist. The tendons of the muscles that cross the wrist are bound down by the *flexor retinaculum* just above the wrist joint. In the same way the tendons are bound down to the fingers to keep them close to the bones.

Muscles of the hand

The hand contains very little muscle, as this would tend to make it clumsy and interfere with its usefulness in grasping and lifting. Many of the muscles that move it are therefore in the forearm (see Fig. 13.14). The hand only contains the short flexor of the thumb, and the adductor and abductor muscles for the digits. These latter are termed the *interosseous muscles*, and are only well developed at

Table 13.4 Muscles of the arm

Name	Position	Origin	Insertion	Action
Biceps (two heads)	The front of the arm	Coracoid process and above the glenoid cavity of the scapula	Radial tuberosity	Flexes elbow and shoulder and supinates the hand
Triceps (three heads)	The back of the arm	One head from the axillary border of the scapula; two from the shaft of the humerus	Olecranon of ulna	Extends elbow and shoulder
Deltoid	Over the shoulder	Acromion and spine of scapula; the clavicle	Deltoid tuberosity of humerus	Abducts the shoulder to a right angle
Brachialis	Crosses the front of the elbow	Humerus	Anterior surface of coronoid process of ulna	Flexes the elbow

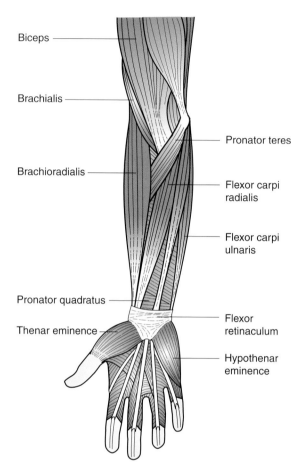

Fig. 13.14 The muscles of the forearm and hand.

the base of the thumb and to a lesser extent at the base of the little finger. Here they form the thenar and hypothenar eminences, and give power to the grip, where adduction of the thumb is particularly important.

MUSCLES OF THE LOWER LIMB

The muscles of the lower limb are much larger and more powerful than those of the upper, as the limb carries the whole weight of the body. They may be divided into the muscles of the thigh, leg and foot.

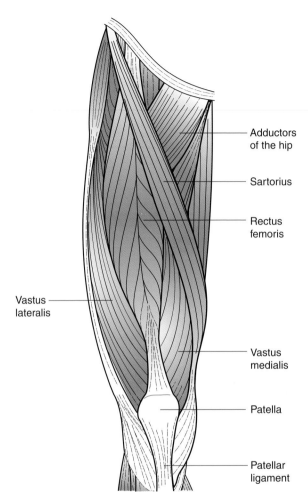

Fig. 13.15 The muscles of the front of the thigh.

Muscles of the thigh

The muscles of the thigh, shown in Figs 13.15 and 13.16, are particularly strong and include:

- The quadriceps femoris
- The hamstrings
- The sartorius
- The adductors of the hip.

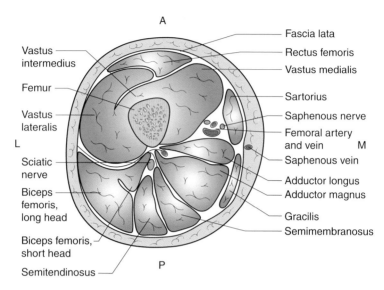

A

Vastus intermedius

Femur

Vastus lateralis

L

Sciatic nerve

Biceps femoris, long head

Biceps femoris, short head

Semitendinosus

Fascia lata

Rectus femoris

Vastus medialis

Sartorius

Saphenous nerve

Femoral artery and vein   M

Saphenous vein

Adductor longus

Adductor magnus

Gracilis

Semimembranosus

P

Fig. 13.16 Section through the thigh, showing the relation of the muscles to the bone and other structures. L, Lateral; M, medial; A, anterior; P, posterior.

The *quadriceps* (so called because it has four heads) is actually a group of four muscles with a common insertion into the patella and, through the patellar ligament, it joins the tibia and is the extensor of the knee joint, used in standing and in the powerful action of kicking. It is made up of the *rectus* or straight muscle, and the three *vastus muscles*: the lateral; intermedial; and medial. Of these the lateral is the longest muscle and lies on the outside of the thigh. The lateral vastus muscle is occasionally used for intramuscular injections as it is situated well away from the blood vessels, nerves and lymphatics of the limb.

The *hamstrings* are the flexor muscles of the knee and are so called because of the strong tendons or 'strings' that they form on either side of the popliteal space, at the back of the knee joint. These can be seen and felt readily when the knee is bent. The hamstring muscles are (Fig. 13.17):

1. The *biceps femoris* is so called because it arises by two heads, one from the ischial tuberosity, one from the back of the femur. It is inserted into the fibula and lies on the outside of the back of the thigh.

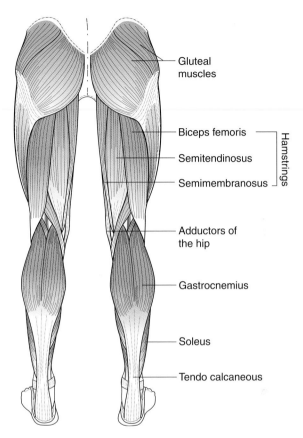

Fig. 13.17   The muscles of the back and lower limbs.

2. The *semitendinosus* is so called because of the length of the tendon by which it is inserted into the tibia. It arises from the ischial tuberosity, with the biceps muscle, and lies in the middle of the back of the thigh.

3. The *semimembranosus* is so called because of the membranous tendon by which it arises from the ischial tuberosity. It is inserted into the tibia and lies on the inner side of the back of the thigh.

   The biceps thus forms the hamstring tendon on the outer side of the back of the knee, and the semitendinosus and semimembranosus form the hamstring tendon on the inner side of the back of the knee. The hamstring muscles are again a very powerful group of

Table 13.5 Muscles of the thigh

Name	Position	Origin	Insertion	Action
Quadriceps femoris	Front of thigh	Ilium and femur	Patella, through which it is joined to the tibia by the patellar ligament	Extension of knee and flexion of hip
Hamstrings	Back of thigh	Ischial tuberosity and femur	Tibia and fibula by tendons at either side of the popliteal space	Flexion of the knee and extension of the hip
Sartorius	Crosses front of thigh	Anterior, superior iliac spine	Tibia at inner side below the knee	Assists in flexion of hip and knee and abduction and external rotation of hip
Adductors of the hip	Inner side of thigh	Pubis and ischium	Linea aspera and medial condyle of femur	Adduct the hip

muscles serving to bend the knee in walking, jumping and climbing, and also, when the tibia is fixed, as in standing, pulling on the ischial tuberosity and helping the gluteal muscles to extend the hip joint.

The *sartorius* or tailor's muscle runs from the superior anterior iliac spine across the front of the thigh to the inner side of the knee, which it crosses, and is inserted into the tibia. It assists in the joint movements that occur when sitting tailor-wise, flexing the hip and knee and rotating the femur.

The *adductor muscles* form the flesh on the inside of the thigh and are muscles by which the hip is adducted, being assisted by the smaller, more superficial muscle, the *gracilis*. These muscles are particularly well developed in horse riders since the rider holds on by the knees, keeping the hips adducted. The adductor muscles run from the pubis and ischium and are inserted into the linea aspera and medial condyle of the femur. The adductor magnus forms the greater part of this muscle and a canal runs through it, which allows the main artery of the thigh to run from the inner side of the thigh to the back of the knee, where it is well protected (Table 13.5).

Muscles of the leg

In the leg there are a few large muscles, shown in Figs 13.18 and 13.19, controlling the ankle and many smaller ones moving the foot. The chief muscles are:

- The gastrocnemius ⎫
- The soleus   ⎬ the calf muscles
- The tibialis anterior
- The flexors and extensors of the digits.

The *gastrocnemius* and *soleus* together form the flesh of the calf, the gastrocnemius lying at the back and the soleus lying in front of it. The gastrocnemius arises by two heads from the femur, forming the borders of the popliteal space below (as the hamstrings do above the knee). The soleus arises from the tibia, not crossing the knee joint and therefore not affecting its movements. Both muscles unite below to form a strong common tendon, the *tendo calcaneus*, by which they are inserted into the calcaneum. The calf muscles raise the heel, causing plantar flexion or, as it is sometimes now called, extension, of the ankle joint, as in walking and running.

The *tibialis* lies in front of the leg just outside the crest of the tibia, where it can be seen and felt when the toes and ball of the foot are

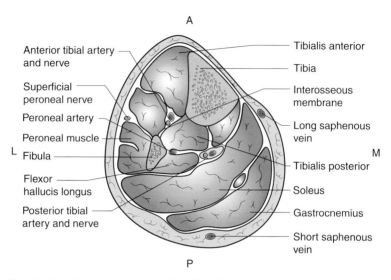

Fig. 13.18 Section through the leg, showing the position of the muscles in relation to the bones. A, Axis of the centre of the limb; M, medial; L, lateral; P, posterior.

raised from the ground. It is this muscle that becomes stiff from abnormal exercise when, for example, people unused to walking up steep slopes begin to climb mountains. It arises from the tibia and fibula below the knee joint, and is inserted into the tarsal and metatarsal bones on the inside of the instep, where its tendon can be seen and felt readily when the ball of the foot is raised from the ground and the sole turned to face in towards the midline. It produces dorsiflexion of the ankle, i.e. bending towards the dorsal surface, or, as it is sometimes now called, flexion, bringing the terminology into line with that used for all other joints. It also inverts the foot (Table 13.6).

Muscles of the foot

The foot, like the hand, contains little muscle – the chief muscles moving the foot lie largely in the leg. The tendons of the extensors of the digits cross the dorsal surface of the foot, the big toe having an individual muscle and tendon. The tendons of the *flexors of the digits* cross the sole and are strong and very important in helping to support the arch of the foot. There is a common flexor for the small toes and a flexor for the big toe. In addition, a short flexor of the toes

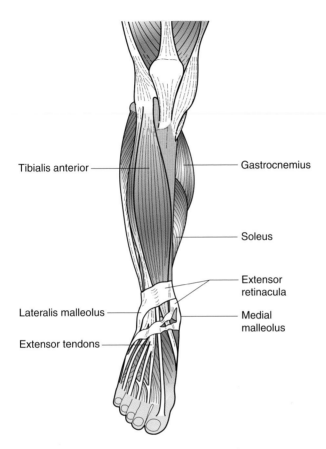

Fig. 13.19   The muscles of the front of the leg and foot.

Table 13.6   Muscles of the leg

Name	Position	Origin	Insertion	Action
Gastrocnemius	The calf	Condyles of the femur	Calcaneum	Plantar flexion of the ankle; raising the heel
Soleus	The calf	Tibia and fibula	Calcaneum	Plantar flexion of the foot
Tibialis anterior	Side of the leg	Tibia	Tarsals and metatarsals inside the instep	Dorsiflexion of the ankle, raising the toes; inversion of the foot

crosses the sole from the calcaneum to the phalanges and also gives support to the arch. Small *interosseous muscles* between the metatarsal bones abduct and adduct the digits but are little used and therefore little developed.

SELF-TEST QUESTIONS

- How are the endomysium, perimysium and epimysium related?
- Describe what is meant by the origin and insertion of a muscle.
- How does skeletal muscle contract?
- What are the groups of muscles of the trunk?
- Which muscles are normally used in respiration?
- Name the muscles of the arm.
- Name the muscles of the leg.

SECTION 4

Internal transport

SECTION INTRODUCTION

Two closely related systems, in both the anatomical and physiological sense, are covered in this section. These are the cardiovascular and respiratory systems and they, respectively, are responsible for circulating blood around the body and delivering oxygen to the blood and removing carbon dioxide. The lymphatic system, which also plays a key role in the defence of the body, is included here because it is responsible for removing excess fluid from the interstices of the peripheral tissues and returning it to the circulation.

The cardiovascular system is composed of the heart and the blood vessels and it is their collective action, with the heart working as a circulatory pump and the blood vessels as a closed transport system, which delivers blood to all the parts of the body. The blood is a tissue with several functions, including the transport of gases, defence against infection, clotting, distribution of heat and transport of nutrients. The gas transport and clotting functions are considered in this section.

Primarily, the respiratory system exchanges oxygen and carbon dioxide between atmospheric air and the blood. This is achieved mechanically in respiratory movements, which draw in atmospheric air and expel air rich in carbon dioxide. Air is mixed in the alveoli to produce alveolar air, and gas exchange between the blood and alveolar air takes place here by diffusion across the thin walls of the pulmonary capillaries and the alveolar membranes. These mechanisms are described.

Chapter 14

The blood

LEARNING OBJECTIVES

After reading this chapter you should understand:

- the composition of blood
- the formation of blood cells
- the formation of red blood cells
- the process of blood clotting
- the functions of blood
- blood groups and the Rhesus (Rh) factor

The circulatory system is the transport system of the body by which food, oxygen, water and all other essentials are carried to the tissue cells and their waste products are carried away. It consists of three parts:

- The *blood* – the fluid in which materials are carried to and from the tissues

- The *heart* – the driving force, which propels the blood
- The *blood vessels* – the routes by which the blood travels to and through the tissues, and back to the heart.

The blood is a thick red fluid; it is bright red in the arteries, where it is oxygenated, and a dark purplish-red in the veins, where it is deoxygenated, having given up some of its oxygen to the tissues – the cause of the colour change – and received waste products taken in from them. It is slightly alkaline; the pH varies very little during life as the cells of the body can live only if the pH is normal. It forms about 5% of the body weight, hence the average volume is 3–4 L.

COMPOSITION OF THE BLOOD

Although apparently fluid, blood actually consists of a fluid and a solid part. When examined under the microscope, large numbers of small round bodies, known as the blood *corpuscles* or cells, can be seen. These form the solid part, while the liquid in which they float is the fluid part or *plasma*. The cells form 45% and the plasma 55% of the total volume.

Plasma

The plasma (the fluid part of the blood) is a clear, straw-coloured, watery fluid similar to the fluid found in an ordinary blister. It is composed of the following:

- Water, which forms over 90% of the whole.
- Mineral salts – these include chlorides, phosphates and carbonates of sodium, potassium and calcium. The chief salt present is sodium chloride. The correct balance of the various salts is necessary for the normal functioning of the body tissues, and there is a total of 0.9% inorganic substances.
- Plasma proteins – albumin, globulin, fibrinogen, prothrombin and heparin.
- Foodstuffs in their simplest forms – glucose, amino acids, fatty acids and glycerol, and vitamins.
- Gases in solution – oxygen, carbon dioxide and nitrogen.
- Waste products from the tissues – urea, uric acid and creatinine.
- Antibodies and antitoxins – these protect the body against bacterial infection.
- Hormones – from the ductless glands.
- Enzymes.

The water in plasma provides fresh water to supply the fluid that bathes all the body cells and renews the water within the cells; 60% of body weight is water and in an average person weighing 70 kg this would be approximately 46 L. Of these 46 L, approximately 29 L are within the cells (intracellular fluid) and 17 L are outside the cells (extracellular fluid). The extracellular fluid is divided between the blood vessels (3 L) and the fluid bathing the cells, called the interstitial fluid (14 L).

The salts in the plasma are necessary for the building of protoplasm and they act as buffer substances neutralizing acids or alkali in the body and maintaining the correct pH of the blood. Blood is always slightly alkaline in health and has a pH of 7.4 (see Chapter 1). In plasma there are approximately 155 mmol/L of positively charged ions, chiefly sodium, balanced by 155 mmol/L of negatively charged ions, mainly chloride and bicarbonate. This is referred to as the *electrolyte balance* and is similar in the *interstitial fluid*. In the intracellular fluid, potassium replaces sodium as the positively charged ions, and phosphate ions and proteins replace chloride as the negatively charged ions.

The proteins that plasma contains give the blood the sticky consistency, called viscosity, which is necessary to prevent too much fluid passing through the capillary walls into the tissues. If there is a deficiency of protein, as in kidney disease, when protein is constantly lost as albumin in the urine, the osmotic pressure of the plasma is lowered and excess fluid escapes into the tissues. This excess fluid is called *oedema*. The viscosity of the blood also assists in the maintenance of the blood pressure. *Albumin* is thought to be formed in the liver and *globulin* is derived from the group of white blood cells called lymphocytes. *Fibrinogen* and *prothrombin* are produced in the liver and are both necessary for the mechanism of the clotting of blood. Plasma without fibrinogen is called *serum*; this can be seen as the yellow fluid that oozes from a cut after a clot has formed. *Heparin* is also formed in the liver and prevents blood clotting in the vessels.

Foodstuffs, in the form of glucose, amino acids, fatty acids and glycerol, are absorbed from the alimentary tract into the blood. They are the endproducts of carbohydrate, protein and fat metabolism.

Urea, uric acid and creatinine are the waste products from protein metabolism. They are made in the liver and are carried by the blood for excretion by the kidneys.

Antibodies and *antitoxins* are complex protein substances providing protection against infection and neutralizing the poisonous bacterial toxins.

Enzymes, which are all protein molecules, produce chemical changes in other substances without themselves entering into the reaction.

The blood cells

The cells are of three types: red blood cells (*erythrocytes*), white blood cells (*leucocytes*) and platelets (*thrombocytes*).

The formation of blood cells

The formation of blood cells takes place in the bone marrow and the mature products are released into the blood stream. Eight different cells are formed and all are formed from one type of pluripotent stem cell, which gives rise to five different lines of cells (Fig. 14.1). The *myeloblast line* gives rise to three types of *granulocyte cells* and the *monoblast* and *lymphoblast* lines give rise to the *agranulocyte* cells. The erythrocytes (red cells) and the *platelets* are formed from their own specific cell lines.

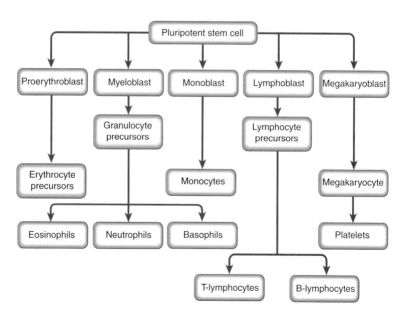

Fig. 14.1 Summary of the formation of blood cells.

Red cells

The red cells (erythrocytes) are disc-shaped bodies, concave on both sides (Fig. 14.2). They are very numerous, numbering about 5 000 000 per cubic millimetre (mm^3) of blood. They have a diameter of 7.2 micrometres (1 micrometre = $1/1000$ mm, abbreviated to μm). They have no nucleus but contain a special protein known as *haemoglobin*. This is a yellow pigment but the massed effect of these numerous yellow bodies is to make the blood appear red. Haemoglobin contains a little iron, which is essential to normal health, though the total amount in the whole body is said to be only sufficient to make a two-inch nail. Haemoglobin has a great affinity for oxygen. As the red cells pass though the lungs the haemoglobin combines with oxygen from the air (forming oxyhaemoglobin) and becomes bright in colour – this makes the oxygenated blood bright red. As the red cells pass through the tissues oxygen is given off from the blood and the haemoglobin becomes a dull colour (reduced haemoglobin) making the blood a dark purplish-red. The haemoglobin is measured in grams per 100 mL; the normal figure is 14–16 g per 100 mL.

The function of the red cells is therefore to carry oxygen to the tissues from the lungs and to carry away some carbon dioxide. This is their sole function, and is dependent on the amount of haemoglobin they contain. If there is a lack of haemoglobin, either because red cells are reduced in number or because each one does not contain the normal quantity of haemoglobin, the individual suffers from *anaemia*.

The red cells are produced in the red bone marrow of spongy bone, which is found in the extremities of long bones and in flat and irregular bones. In childhood the red bone marrow also extends

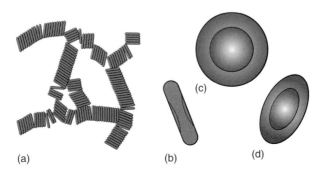

Fig. 14.2   Red cells. (a) In rouleaux, as seen in shed blood under the microscope. (b)–(d) Three views of a single cell.

throughout the shaft of the long bones, as children have a greater need for the production of red cells.

Red cells pass through various stages of development in the bone marrow. *Erythroblasts* are large cells containing nuclei and a small quantity of haemoglobin. These develop into *normoblasts*, which are smaller cells with more haemoglobin and smaller nuclei. The nucleus then disintegrates and disappears, and the cytoplasm contains fine threads, at which stage the cells are called *reticulocytes*. Finally, the threads disappear and the fully mature erythrocyte passes into the blood stream. In health, almost all red cells in the blood should be erythrocytes, with only an occasional reticulocyte. Many factors are necessary for the normal formation of red blood cells. The overproduction of red cells, which may occur at high altitudes, is called polycythaemia.

Protein is required for the manufacture of protoplasm.

Iron is needed for the haemoglobin. Very little iron is excreted. As the red cells are broken down the iron is stored and used again but a certain amount of iron must be taken in the diet. A man requires about 10 mg of iron per day but women require about 15 mg to replenish the menstrual loss and the depletion of iron reserves that occur during pregnancy, labour and lactation. Iron-containing foods include red meat, egg yolk, green vegetables (such as lettuce and cabbage), peas, beans and lentils.

Vitamin B_{12} (cyanocobalamin) is necessary for the maturation of red blood cells and is usually found in adequate quantities in the diet in temperate climates. It can be absorbed from the small intestine only when it has been combined with the *intrinsic factor*, which is secreted by the stomach. Together these two substances are known as the anti-anaemic factor (or haemopoietic factor), which is stored in the liver and passed to the bone marrow as necessary. Vitamin B_{12} is also known as the extrinsic factor.

Other necessary factors, even if only in small quantities, are vitamin C, folic acid (one of the vitamin B complex), the hormone thyroxine and traces of copper and manganese.

Red blood cells live in the circulation for about 120 days after which they are ingested by the cells of the *monocyte/macrophage system* in the spleen and lymph nodes. Here the haemoglobin is broken down into its component parts, which are carried to the liver. The *globin* is returned to the protein stores or is excreted in the urine after further breaking down. The *haem* is further split into iron, which is stored and used again, and pigment, which is converted by the liver into bile pigments and is excreted in the faeces. Red cell

 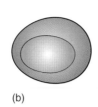 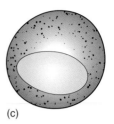

(a)   (b)   (c)

Fig. 14.3   White cells. (a) A polymorphonuclear leucoyte. (b) A lymphocyte. (c) A monocyte.

production and breakdown usually proceed at the same rate so that the number of cells remains constant.

White cells

The white cells (leucocytes) are larger than the red cells, measuring about 10 μm in diameter, and they are less numerous. There are $7–10 \times 10^9$ per litre of blood though this increases considerably to 30×10^9 per litre when infection is present. This increase is known as *leucocytosis*. The leucocytes are of three different types (Fig. 14.3), as follows.

Polymorphonuclear leucocytes   Polymorphonuclear leucocytes are also known as *granulocytes* because of the granular appearance of the cytoplasm. The nucleus gradually develops several lobes, hence the name (poly = many; morph = form). These cells make up about 75% of the total white cells. They are made in the red bone marrow and survive for about 21 days. Granulocytes can be further classified according to their staining properties:

1. *Neutrophils* (70%) absorb both acid and alkaline dyes. They are able to ingest small particles, e.g. bacteria and cell debris. This ability is called *phagocytosis* and they are sometimes known as *phagocytes*. A decrease in the production of these cells is called agranulocytosis and this may occur as a result of exposure to radiation. They have amoeboid movements and can pass out of the blood stream through the capillary walls to accumulate where there is infection.

2. *Eosinophils* (4%) absorb acid dyes and stain red. An increase in their number occurs during allergic states such as asthma, and during infestation with worms.

3. *Basophils* (1% or less) absorb alkaline or basic dyes and stain blue. They contain heparin and histamine.

Lymphocytes   Lymphocytes make up about 20% of the total white cell count. They are made in the lymph nodes and in the lymphatic tissue, which is present in the spleen, liver and other organs. They show some amoeboid movement but are not actively phagocytic. They are concerned with the production of antibodies.

Monocytes   Monocytes make up about 5% of the total white cell count. They are the largest of the white blood cells and have a horse-shoe-shaped nucleus. They show both amoeboid movement and are phagocytic, and are part of the monocyte/macrophage system (see p. 212).

An overproduction of immature leucocytes leads to the condition of *leukaemia*. There are several types of leukaemia, all are fatal if not treated and some forms are more responsive to treatment, usually by chemotherapy, than others.

Platelets

Platelets (thrombocytes) are even smaller than red blood cells and are made in the bone marrow. There are about 250 $\times$ 10^9 per litre of blood. They are necessary for the clotting of blood.

THE CLOTTING OF BLOOD

The process whereby blood loss from the body is prevented following a cut is called haemostasis and involves three stages, which work together:

- *Vascular spasm* – narrowing of the lumen of the cut blood vessels to slow down the loss of blood
- *Formation of a platelet plug* – to stop the flow of blood from the cut
- *Clotting of fibrin* around the plug and retraction of fibrin – to seal the cut and pull its edges together.

The *clotting process* (Fig. 14.4) is very complex and involves many factors. The endpoint in the process is the formation of an insoluble fibrin clot from soluble fibrinogen and this process is stimulated by the formation of *thrombin*. The formation of thrombin is stimulated by the formation of *prothrombin activator* and there are two systems whereby this is achieved – the extrinsic and the intrinsic systems.

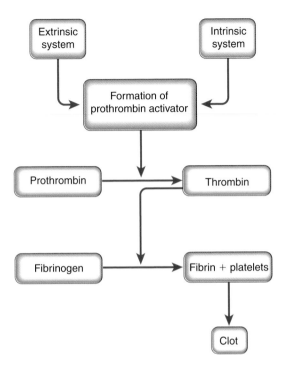

Fig. 14.4   The clotting process.

The *extrinsic system* is stimulated by damaged tissue and quickly forms a very small amount of fibrin to form a clot. The *intrinsic system* takes a few minutes to work but leads to the formation of a relatively large amount of fibrin to complete the formation of the *clot*. As soon as the clot is formed it is broken down by the action of an enzyme called plasmin. This allows the clot to be removed in order to begin the process of wound healing (see Chapter 24).

Factors affecting clotting

Prothrombin is made in the liver and vitamin K is necessary for its manufacture. Vitamin K is present in green vegetables (e.g. lettuce and cabbage) and is also manufactured in the intestines by bacterial action. It can be absorbed from the intestines into the blood only in the presence of bile. If bile is not present, as in some forms of *jaundice*, prothrombin may be lacking and the tendency to bleed is increased. There are a group of genetic disorders known as the haemophilias in which individuals fail to produce particular protein factors essential for clotting. These disorders lead to prolonged bleeding and also bleeding in deep tissues in the body from very minor injuries. These conditions can only be treated by replacing the missing clotting factors.

Heparin is a protein normally present in blood and is formed in the liver. It prevents blood clotting in the vessels and is therefore an anticoagulant.

THE FUNCTIONS OF BLOOD

The functions of blood are to:

- Carry nutrients to the tissues
- Carry oxygen to the tissues in oxyhaemoglobin
- Carry water to the tissues
- Carry away waste products to the organs that excrete them
- Fight bacterial infection through the white cells and antibodies
- Provide the materials from which glands make their secretions
- Distribute the secretions of ductless glands and enzymes
- Distribute heat evenly throughout the body and so regulate the body temperature
- Arrest haemorrhage through clotting.

BLOOD GROUPS

Blood from one individual cannot always be safely mixed with that of another. This fact became evident with the introduction of *blood transfusion*, which at first sometimes cured but sometimes killed the patient. This was due to the fact that blood is of four basic groups. If blood of differing groups is mixed, the red corpuscles may become sticky and form clumps. This is termed *agglutination* and is fatal as the clumps of red cells block the blood vessels and obstruct the circulation, and the kidneys are severely damaged by excretion of excessive quantities of pigment from destroyed red cells.

When a blood transfusion is required, it is necessary first to find which group the person belongs to and then to find a donor belonging to the same group. Blood groups are named according to the presence or absence of substances called *agglutinogens*, which are present in the red blood cells. There are two agglutinogens: A and B. If A is present the blood group is called Group A; if B, Group B. If both A and B are present the blood group is called Group AB, and if neither is present it is Group O. Having found to which group the patient belongs (Table 14.1) and found blood donated by a person

Table 14.1   Blood groups

Blood group	Agglutinogen in red cells	Agglutinin in plasma	Transfusion possible
A	A	Anti-B	Groups A and O
B	B	Anti-A	Groups B and O
AB	A and B	Neither	Any group
O	None	Anti-A and anti-B	Group O only

belonging to the same basic group, a sample of red blood cells from the donor's blood is mixed with some plasma from the patient who is to receive the transfusion (now called the recipient). This is because plasma contains substances called *agglutinins*, which cause agglutination of the red cells if incompatible blood groups are mixed. Agglutinins are called anti-A and anti-B, and plasma contains all those agglutinins that will not affect its own red cells. Therefore the plasma of Group A contains anti-B agglutinin, the plasma of Group B contains anti-A agglutinin, the plasma of group AB contains no agglutinins, and the plasma of Group O contains both anti-A and anti-B agglutinins. When the donor's red cells are mixed with the recipient's plasma in the laboratory, it can be seen under the microscope whether agglutination occurs. It will be noticed that Group AB has no agglutinins in the plasma and therefore cannot cause any red cells to agglutinate – this means that a patient with blood belonging to this group will probably be able to receive blood from any other group and the group is therefore known as the *universal recipient*. Group O contains no agglutinogens in the red cells and therefore they cannot be made to agglutinate by the agglutinins in any plasma. Blood belonging to this group can therefore be given to a patient belonging to any group and it is known as the *universal donor*. In practice the compatibility of blood is always checked very carefully before it is given to a patient.

Rhesus factor

In addition to the ABO grouping there is an additional factor present in the blood of about 85% of the population. It is an agglutinogen called the Rhesus factor. Those who possess this factor are called Rhesus positive (Rh+) and the remaining 15% are Rhesus negative (Rh−). If a Rh− person receives the blood of a Rh+ donor the agglutinogen stimulates the production of anti-Rh agglutinins called anti-D. If a second Rh+ transfusion were given later the transfused cells would be agglutinated and destroyed (haemolysed) with serious or fatal results to the recipient. This factor can also cause difficulty during pregnancy. If a Rh− mother is carrying a Rh+ fetus the mother may begin to produce anti-Rh agglutinins, which may then destroy the baby's red cells. The baby may overcome this spontaneously or may require exchange transfusion.

SELF-TEST QUESTIONS

- What is the major constituent of blood?
- How many different types of blood cell are there?
- Describe an erythrocyte.
- Compare and contrast the intrinsic and extrinsic systems of blood clotting.
- What is the blood group of a universal donor and a universal recipient?

Chapter 15

The heart and blood vessels

CHAPTER CONTENTS

LEARNING OBJECTIVES

After reading this chapter you should understand:

- the structure of the heart and major blood vessels
- the function of the valves of the heart
- the function of the heart
- the conduction system of the heart
- control of the heart
- the structure of the blood vessels
- the circulation and how it works
- the control of blood pressure

The purpose of this chapter is to consider the anatomical location and structure of the heart and major blood vessels. Some aspects of the function of the heart are included such as the cardiac cycle, the pathway of blood around the heart and the generation of electrical activity at the sinoatrial node.

THE HEART

The heart is a hollow, muscular, cone-shaped organ (Fig. 15.1). It lies between the lungs in an area called the mediastinum, behind the body of the sternum with two-thirds of its bulk on the left side. The circular base of the cone is directed upwards and to the right and the *apex* points downwards, forwards and to the left. It is usually on the level of the fifth intercostal space about 9 cm from the midline. The heart measures about 12 cm from base to apex and about 9 cm in width; it is about 6 cm thick.

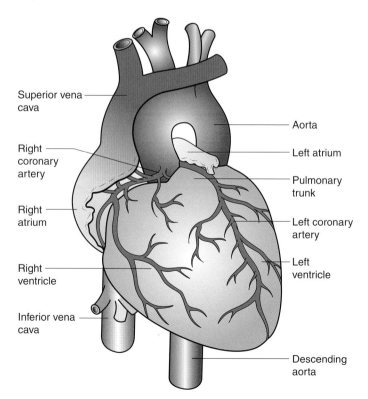

Fig. 15.1 Front view of the heart.

General structure

The heart is divided from base to apex by a muscular partition called the *septum*. The two sides of the heart have no communication with each other in health. Additionally, each side is subdivided into an upper and lower chamber. The upper chamber on each side, the *atrium*, is smaller and is a receiving chamber into which the blood flows through veins. The lower chamber, the *ventricle*, is the discharging chamber from which the blood is driven into the arteries. Each atrium communicates with the ventricle below it on the same side of the heart through an opening, guarded by a valve called the *atrioventricular* valve (Fig. 15.2).

The heart is composed of cardiac muscle, the *myocardium*, on the action of which the circulation of the blood depends. The myocardium varies in thickness, being thickest in the left ventricle, thinner in the right ventricle and thinnest in the atria.

The atria and ventricles are lined with a thin, smooth, glistening membrane called the *endocardium*, which consists of a single layer of endothelial cells and which is continuous with the valves and the lining of the blood vessels.

The *pericardium* covers the heart and the roots of the great vessels and has two layers. The outer layer of fibrous pericardium is securely

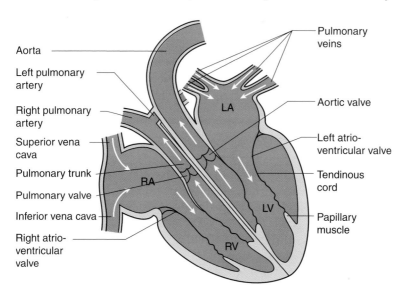

Fig. 15.2 Diagram showing the flow of blood through the heart. LA, Left atrium; LV, left ventricle; RA, right atrium; RV, right ventricle.

anchored to the diaphragm, the outer coats of the great vessels and the posterior surface of the sternum and therefore maintains the heart in its position in the chest. Because of its fibrous nature it also prevents overdistension of the heart. The inner layer, the serous pericardium, lines the fibrous pericardium and is invaginated by the heart; it therefore also has two layers. The inner layer is known as the visceral portion, or *epicardium*, and it is reflected back to form the outer or parietal layer. The layers are normally all in close contact and are moistened by fluid that exudes from the serous membrane; this prevents any friction as the heart continually contracts and relaxes. In inflammatory conditions, such as *pericarditis*, the amount of fluid within the pericardium may embarrass the action of the heart and may be aspirated.

Valves

The heart is provided with valves to prevent the blood from flowing in the wrong direction. There are four main valves.

The *right atrioventricular* (or tricuspid) valve lies between the right atrium and the right ventricle. It consists of three triangular flaps or cusps, each consisting of a double layer of endocardium strengthened with fibrous tissue. The undersurface of the cusps provides attachment for a number of fine, tendinous cords, called the chordae tendinae, which originate in the papillary muscles in the wall of the ventricle. When the ventricle contracts, the blood is pushed back towards the atrioventricular opening but is prevented from entering the atrium by the cusps of the valve, which close because of the increased pressure in the ventricle. The contraction of the papillary muscles exerts tension on the chordae tendinae, which prevent the valve cusps from being carried into the right atrium (Figs 15.3 and 15.4).

The *left atrioventricular valve* is also called the mitral valve (due to its resemblance to a bishop's mitre) (Fig. 15.5). The structure is similar to that of the right atrioventricular valve. It prevents the backflow of blood into the left atrium during contraction of the left ventricle.

The *aortic valve* consists of three cusps, which surround the entrance into the aorta from the left ventricle (Figs 15.6 and 15.7). The cusps are half-moon shaped and are fixed by their curved edges to the wall of the aorta, the straight edge being free so that pockets are formed facing into the aorta. As the blood flows from the left ventricle into the aorta the cusps lie flat against the vessel wall; as the ventricle relaxes the pockets fill with blood and bulge out, meeting in the centre and blocking the opening completely, thus preventing blood

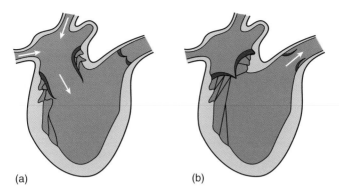

Fig. 15.3 The action of the valves of the heart. (a) The position of the valves during contraction of the atrium and relaxation of the ventricle. (b) The position of the valves during contraction of the ventricle and relaxation of the atrium.

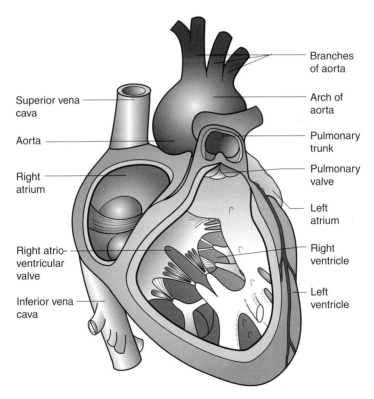

Fig. 15.4 Inside the right side of the heart.

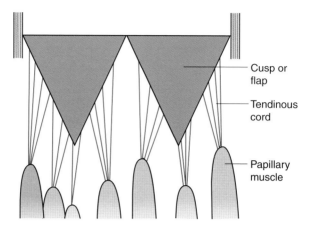

Fig. 15.5 Diagrammatic representation of the mitral valve laid open.

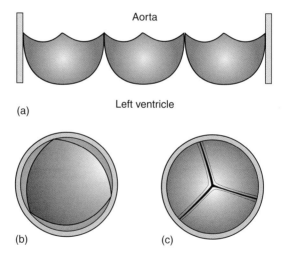

Fig. 15.6 Diagrammatic representation of the aortic valve. (a) Laid open; (b) seen from above; (c) closed.

flowing back into the ventricle. The coronary arteries, which supply the heart muscle with oxygenated blood, arise from the aorta just above the attached edges of the cusps of the aortic valve.

The *pulmonary valve* guards the opening from the right ventricle into the pulmonary trunk. It is similar in structure and action to the aortic valve.

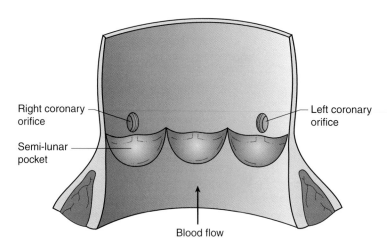

Fig. 15.7   The aortic valve.

The blood that returns to the heart from the myocardium passes through the *coronary sinus* and pours directly into the right atrium. The opening of the coronary sinus is protected by a thin, semicircular valve, called the valve of the coronary sinus, which prevents backflow of blood into the sinus during contraction of the right atrium. There is also an imperfect valve guarding the opening from the inferior vena cava into the right atrium; it is called the valve of the inferior vena cava.

The blood supply to the heart muscle is by the right and left *coronary arteries*, which are the first branches from the aorta. There are a number of places within the heart muscle where these two arteries join, or anastomose, but most of the blood returns from the myocardium in veins that empty into the coronary sinus. Blockage of the left coronary artery by a blood clot leads to a reduction in the supply of blood to the left ventricles: this is a myocardial infarction. Depending upon where the infarction takes place the outcome is more or less serious. An infarction near the point where the left coronary artery leaves the aorta can be fatal.

Function

The heart is a pump whose purpose is to drive the blood into and through the arteries, but the right and left sides of the heart function quite separately from one another.

Blood from all parts of the body is returned to the right atrium through the two large veins, the superior and inferior *venae cavae*. When it is full the right atrium contracts and drives the blood through the right atrioventricular valve into the right ventricle, which, in turn, contracts sending the blood through the pulmonary valve and into the *pulmonary trunk*. The pulmonary trunk divides into right and left *pulmonary arteries*, which carry the blood to the lungs where gaseous exchange occurs. The blood is finally collected into four *pulmonary veins*, which return the blood to the left atrium. When it is full the left atrium contracts, simultaneously with the right atrium, and the blood is driven through the left atrioventricular valve into the left ventricle. This chamber contracts, simultaneously with the right ventricle, and sends the blood into the aorta, the main artery of the body.

The heart contracts about 70 to 80 times each minute throughout life, though the rate varies with age, emotion, exercise and other influences. Each beat is a cycle of events that lasts about 0.8 seconds (Fig. 15.8).

Blood pours into the atria from the great veins until both are full and they then contract simultaneously, emptying their contents into

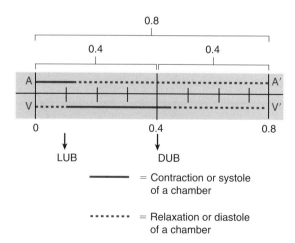

Fig. 15.8 Graph to show the sequence of events in the cardiac cycle. A–A', the contraction and relaxation of the atrium; V–V', the contraction and relaxation of the ventricle. The cycle occupies 0.8 seconds, of which 0.4 seconds is occupied by the contraction of both atria and ventricles, and the remaining 0.4 seconds by the relaxation of all chambers of the heart. (Note 0.8 seconds is the time of one beat if the pulse rate is 75 per minute.)

the ventricles. Atrial contraction lasts about 0.1 seconds. Rising pressure in the ventricle as blood enters forces the atrioventricular valves to close and causes the first heart sound, which can be heard through a stethoscope placed over the apex of the heart and has been likened to the word 'lub'. Ventricular contraction follows, lasting 0.3 seconds, and forcing open the pulmonary and aortic valves. Blood is forced into the aorta and pulmonary trunk and as the ventricles relax the pressure in the great vessels forces the aortic and pulmonary valves to close and causes the second heart sound, which can be heard best over the second right rib and has been likened to the word 'dub' because it is a sharper sound. During ventricular contraction the atria are relaxed. Following ventricular contraction the whole heart is relaxed for approximately 0.4 seconds. The word for contraction is *systole* and that for relaxation is *diastole*.

Conducting mechanism

Cardiac muscle, unlike skeletal muscle, has the property of being able to contract rhythmically independent of any nerve supply. The autonomic nervous system modifies the speed of the heart beat but the heart could be removed from an animal and continue to beat. The impulse to contract arises through spontaneous depolarization in an area of specialized tissue near the entry of the superior vena cava into the right atrium – called the *sinoatrial node* – the *pacemaker* of the heart. The impulse then passes across both atria in more or less concentric rings, made possible by the fact that cardiac muscle fibres are branching, to the atrioventricular node situated in the septum near the junction of the atria and ventricles. After a slight pause the impulse spreads down the *atrioventricular bundle*, which is in two strands, one in the right ventricle and one in the left. The strands break up into special fibres, called *Purkinje fibres*, from which branches pass beneath the endocardium to all parts of the ventricles.

The sinoatrial node has the fastest rhythm, about 70–74 beats per minute, and it imposes this rate on the other areas of conduction. The ventricles are able to contract independently of the atria and they will do this if the conducting mechanism is affected by disease, but they will contract much more slowly (about 40 beats per minute). This condition is known as *heart block* and may be serious as the tissues are unlikely to receive an adequate blood supply. In other cases, some impulses pass down under the atrioventricular bundles but some do not, so the ventricles contract once for every two or three atrial contractions; this condition is known as partial heart block.

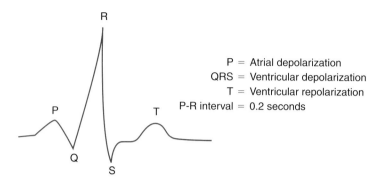

Fig. 15.9   The electrocardiogram.

The electrocardiogram

If electrodes are placed on the body and attached to an instrument called an electrocardiograph, it is possible to detect the electrical activity in the heart. The resulting pattern, which can be observed on an oscilloscope or traced onto paper, is known as the electrocardiogram. This trace is helpful to physicians in allowing the detection of abnormalities, such as myocardial infarction, which alter the electrical activity in the heart. The trace is characteristic and has three phases (Fig. 15.9). The P wave represents the depolarization of the atria and the QRS complex, which masks the repolarization (the return to the resting potential difference across the heart cell membranes), represents the depolarization of the ventricles. Finally, the T wave represents the repolarization of the ventricles.

Nervous control

The heart is innervated by the autonomic nervous system. The *vagus nerve* (tenth cranial nerve) slows the heart rate and causes decreased power of contraction by conveying impulses to the sinoatrial node. The sympathetic nerves speed the heart rate and increase the force of contraction (see Chapter 5). This dual innervation of the heart is co-ordinated by the cardiac centre in the medulla oblongata in the brain.

The heart rate is also controlled by reflex by two sets of receptors. Pressure receptors (*baroreceptors*) are sensitive to changes in blood pressure. They are found in the *carotid arteries* and the *arch of the aorta*. If the blood pressure increases, less sympathetic stimulation and more parasympathetic stimulation is observed and the heart

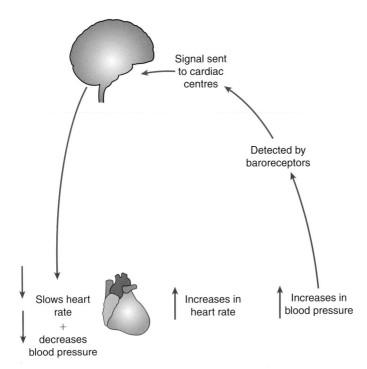

Signal sent
to cardiac
centres

Detected by
baroreceptors

Slows heart
rate
+
decreases
blood pressure

Increases in
heart rate

Increases in
blood pressure

Fig. 15.10   Negative feedback loop controlling heart rate and blood pressure.

rate slows, thus helping to lower the blood pressure. This is an example of a homeostatic mechanism that works by negative feedback (Fig. 15.10). *Chemoreceptors* are sensitive to the amount of carbon dioxide present in the blood. They are found in the neck near the carotid arteries and near the aorta and are sensitive to lack of oxygen. Impulses are conveyed to the cardiac centre and the heart rate is accelerated to increase the blood supply, and therefore the oxygen supply, to the tissues.

In health the heart rate varies considerably, for example:

- Rest slows the heart rate and exercise quickens it
- Increasing age slows the heart rate – infants have a pulse rate of 120–140 at birth, which slows throughout life and into old age
- Women have a slightly faster pulse rate than men
- Emotion and excitement speed the pulse rate.

In disease, conditions such as fever, haemorrhage, shock and hyperthyroidism speed the heart rate, while increased pressure on the brain and heart block slow it. Drugs can also be used to speed or slow the pulse rate.

THE BLOOD VESSELS

Contraction of the ventricles sends the blood to all parts of the body through a complicated series of tubes called *arteries*, which branch into tiny vessels called *arterioles*. These in turn are continuous with a network of microscopic vessels named *capillaries*. Blood is then collected into tiny vessels called *venules*, which unite to form *veins*. These join one another, forming large veins, which finally return the blood to the heart. With age the arteries tend to become thicker or 'hardened' and this is known as arteriosclerosis. In addition, fatty plaques are laid down in the arteries (a condition called atheroma) and when these plaques co-exist with arteriosclerosis the condition is known as atherosclerosis. In some people, these processes become very advanced and they contribute towards circulatory disorders such as myocardial infarction. Atheroma, in particular, is associated with cigarette smoking.

Structure

Arteries are thick-walled vessels and, with one exception, they carry oxygenated blood. The exception is the pulmonary trunk, which divides into two pulmonary arteries and carries deoxygenated blood from the right ventricle to the lungs. All arteries have three layers (Fig. 15.11):

- The outer coat, or *tunica adventitia*, which consists of collagenous and elastic fibres
- The middle coat, or *tunica media*, which consists mainly of smooth muscle with elastic fibres and some collagen fibres

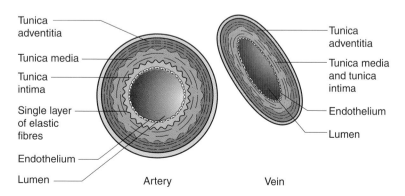

Fig. 15.11   The structure of arteries (left) and veins (right).

- The lining, or *tunica intima*, which consists of a layer of endothelial cells and provides a smooth surface over which the blood can flow without clotting.

All parts of the body must have a blood supply and the arteries are no exception. Very tiny blood vessels bring a blood supply to the arterial walls and equally small venules return the blood to the veins. Lymph vessels and nerve fibres are also present.

Arterioles have the same three structures as arteries but the intima and media are thin and the adventitia is relatively thicker than the adventitia of an artery. There are also more muscle fibres and fewer elastic fibres.

Capillaries form a network between arterioles and venules (Fig. 15.12). They consist of a single layer of endothelial cells similar to those that line all other blood vessels (Fig. 15.13).

Venules and veins both have three coats, as do arteries, but the middle coat of a vein is much thinner than that of an artery. Many veins are provided with valves to prevent blood flowing in the wrong direction. Each valve is formed of a double layer of endothelium

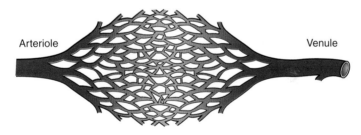

Arteriole   Venule

Fig. 15.12   Diagram of a capillary network.

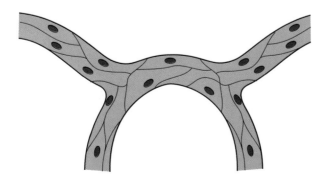

Fig. 15.13   Capillary network, showing the wall of a simple endothelium.

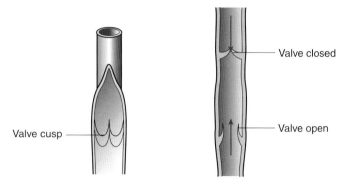

Valve closed

Valve cusp

Valve open

Fig. 15.14 The valves in the veins.

strengthened by connective tissue and elastic fibres. The valve cusps are semi-lunar and are attached by their convex edges to the wall of the vein (Fig. 15.14). Veins, like arteries, have their own blood supply and nerve fibres though the latter are not so numerous.

THE MECHANISM OF CIRCULATION

Pulse

Blood leaving the left ventricle of the heart is rich in oxygen and bright red. It is pumped into the aorta by contraction of the left ventricle, which initiates an area of increased pressure that travels along the arteries rather like a wave. When the blood is pumped out of the left ventricle the aorta is already full, so it must distend to accommodate the additional blood. As the left ventricle relaxes, the aortic valve closes and the elastic aorta recoils to its original diameter. This recoil of the aorta is very important because it is the mechanism by which the blood is continuously driven round the body even when the ventricle is relaxed. The distension and recoil of the aorta sets up a wave of distension and recoil called the *pulse*, which travels along all the large arteries and which can be felt with the fingers wherever an artery can be compressed gently against a bone. Since the heart beat produces the pulsation, the rate and character of the beat can be judged by taking the pulse.

Blood pressure

The blood pressure is the force that the blood exerts on the walls of the blood vessels. It varies in the different blood vessels and also with the heart beat. The pressure is greatest in the large arteries leaving the heart, and gradually falls in the arterioles until, when it reaches the capillaries, it is so slight that the least pressure from without will obliterate these vessels and drive the blood out of them. This can be seen by pressing lightly on the nail or letting a piece of glass rest lightly on the skin. (It is for this reason that it is so important to change the position of a patient confined to bed frequently as the tissue carrying the body's weight has little blood circulating through it.) In the veins the pressure is lower still until ultimately in the big veins approaching the heart there is 'suction', i.e. a negative pressure instead of a positive one, on account of the 'suction' exerted by the heart as its chambers relax.

The pressure in the large arteries varies with the heart beat. It is highest when the ventricle contracts (this is called *systolic pressure*) and lowest when the ventricle relaxes (*diastolic pressure*).

Measurement of blood pressure

The pressure of the blood is measured by the weight of the column of mercury (Hg) that it will support. The height is calculated in millimetres (hence mmHg). The normal arterial pressure is 110–120 mmHg systolic pressure and 65–75 mmHg diastolic pressure. The apparatus for measuring blood pressure is called a *sphygmomanometer*. There are different types but the most common consists of a hollow rubber cuff encased in material that is fixed evenly round the limb without constricting it in any way. To this is attached a small hand pump, by which air can be pumped into the cuff, expanding it and compressing the limb and the blood vessels it contains. The air cushion is also linked to a manometer so that the pressure of the air within it can be read and recorded.

To measure blood pressure the air cuff is fixed round the arm and the pulse at the wrist is felt with one hand while the other is used to inflate the cuff to a pressure above the point at which the radial pulse can no longer be felt. A *stethoscope* is placed over the brachial artery pulse in the cubital fossa and the pressure in the cuff is slowly lowered by means of the release valve. As the pressure falls no sounds are heard until the systolic blood pressure is reached, at which point a tapping sound is heard in the stethoscope. At this point the reading on the *manometer* must be noted. As the pressure in the cuff is further reduced the sounds become louder, until the diastolic blood pressure is reached, at which point the sound is altered in character and becomes muffled. A little further decrease in cuff pressure causes the sound to disappear completely. The diastolic blood pressure is noted as the point at which the character of the sound changes. These are the principles of blood pressure measurement. However, after fixing the cuff, the remainder is usually done by automated equipment in hospital.

Arterial pressure

The arterial pressure is maintained by five factors:

- Cardiac output
- Peripheral resistance
- Total blood volume
- Viscosity of the blood
- Elasticity of the arterial walls.

The *cardiac output* is the amount of blood pumped out per minute. When the left ventricle contracts, about 70 mL of blood is pushed into the aorta, which is already full, causing it to distend. This is the *stroke volume*.

The *peripheral resistance* is the resistance offered by the small blood vessels, particularly the arterioles, to the flow of blood. This resistance prevents blood flowing too quickly into the capillaries and in this way helps to maintain the blood pressure in the arteries. The lumen of the arterioles can be altered by the action of vasomotor nerves and by epinephrine and norepinephrine from the adrenal glands. If the lumen is narrowed the resistance to blood flow is increased and arterial blood pressure is raised. If the lumen is enlarged, more blood will pass through quickly to the capillaries and the blood pressure will be lowered.

The *blood volume* is the total amount of circulating blood. If this is reduced by the loss of whole blood, as in haemorrhage, or by loss of fluid from the circulation, as in shock, burns or dehydration, the blood pressure will be lowered.

The *viscosity* of the blood is its stickiness. Blood is two or three times more viscous than water. The viscosity depends partly on the plasma, particularly the plasma proteins, and partly on the number of red cells. The viscosity of the blood will be reduced if large quantities of intravenous normal saline are given, and lowered viscosity is associated with lowered blood pressure.

The *elasticity* of the arterial walls allows distension of the aorta when the ventricle contracts and elastic recoil as the ventricle relaxes. This recoil pushes the blood on and maintains the diastolic pressure of the blood. Distension and recoil occur throughout the arterial system. Elasticity may be reduced in atheroma, which is a degenerative disease of the arteries. In this condition the blood pressure will be raised because of loss of elasticity. In practice, the blood volume, the blood viscosity and the elasticity of the arterial walls do not change. The blood pressure (BP) can be explained simply in terms of the cardiac output (CO) and the peripheral resistance (PR):

$$BP = CO \times PR$$

High blood pressure

A high blood pressure is a systolic pressure of at least 150–180 mmHg. The diastolic pressure is also raised as a general rule, and a high diastolic pressure, e.g. at least 90–120 mmHg, is injurious as it is a strain on the heart. A high blood pressure is dangerous in that the raised pressure may cause rupture of a blood vessel. Rupture is particularly likely to occur in the brain and is one of the causes of *cerebrovascular accident* or stroke. There is also a strain on the heart, which may result in heart failure.

Low blood pressure

A low blood pressure is a systolic pressure of 100 mmHg or less. It occurs in cases of haemorrhage, shock and collapse, heart failure, and in disease of the adrenal glands. The danger lies in an insufficient supply of blood to the vital centres of the brain. The treatment is therefore:

1. To place the patient flat, with the foot of the bed raised if necessary, to send blood by gravity to the vital centres of the brain.

2. To give heart stimulants and circulatory stimulants such as epinephrine and norepinephrine to contract the blood vessels.

3. To give fluids such as saline by intravenous, subcutaneous or rectal infusion, or blood or plasma transfusion, to increase the fluid in circulation.

As already mentioned, blood vessels have their own nervous supply. Arterioles are supplied with *vasomotor nerves* through which they can be dilated or contracted according to the needs of the tissues. The muscular walls of the arterioles are also affected by hormones from the adrenal glands, particularly epinephrine and norepinephrine, which cause contraction of the muscle fibres and consequent narrowing of the lumen of the arteriole. Arterioles do not pulsate but can dilate to carry more blood to an organ that is working and can contract so that they carry less blood when an organ is resting. In this way arterioles control the distribution of blood to the various organs of the body and are important in the maintenance of blood pressure as their contraction offers resistance to the outflow of blood from the large elastic arteries.

Capillaries receive blood from the arterioles and pass it to the venules. Because their walls are only one cell in thickness, oxygen, water and food substances in solution are able to pass from the blood through the walls to supply the tissue cells, and waste products can pass back from the tissues to be carried away by the blood.

Capillaries form a network with considerable numbers of them forming *anastomoses* so that varying amounts of blood may be brought to the part as required.

Venous return

Blood is collected from the capillary network into venules and then into veins. By the time it reaches the veins it has given up most of its oxygen and is a dark purplish colour. Valves are necessary in the veins to prevent the blood flowing in the wrong direction.

Venous return depends on three factors:

- Suction as the atria relax
- Suction during inspiration – this draws blood towards the heart as well as drawing air into the lungs
- Pressure on the thin-walled veins by contraction of muscles – as the lumen of the vein is reduced in size, blood would be forced in both directions were it not for the pressure of valves, which allow the blood to flow towards the heart only.

The force of 'suction' is the most important factor in the maintenance of venous pressure. If the force is lessened, as in heart failure, venous return is impaired and congestion occurs.

The blood pressure throughout the cardiovascular system is shown in Fig. 15.15. The blood pressure is highest in the elastic arteries nearest the heart and lowest near the right side of the heart in the vena cava.

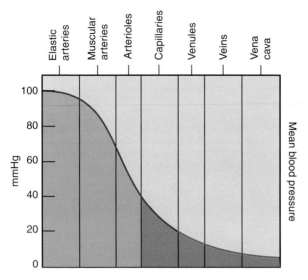

Fig. 15.15 Mean blood pressure.

SELF-TEST QUESTIONS

- Describe the pathway of blood through the heart.
- What are the heart sounds?
- Describe the electrocardiograph.
- Explain the control of the heart through changes in blood pressure.

Chapter **16**

The circulation

LEARNING OBJECTIVES

After reading this chapter you should understand:

- the systemic, pulmonary and hepatic portal circulations
- the major branches of the aorta
- the tributaries of the vena cava

The circulation of the blood is divided into three main parts:

- The systemic circulation
- The pulmonary circulation
- The portal circulation.

THE SYSTEMIC CIRCULATION

The vessels carrying the blood from the left ventricle through the body generally to the right atrium are those that constitute the *systemic circulation*. Arteries may subdivide into several branches at the same point or several branches may be given off in succession. Arteries do not always end in capillaries but may unite with one another, forming anastomoses. Examples are found in the brain where two vertebral arteries anastomose to form the basilar artery, and where two anterior cerebral arteries are connected by the anterior communicating artery (see Fig. 16.2). An enlarged *anastomosis* may provide a *collateral circulation* if a vessel is occluded by accident or disease; sudden occlusion may be followed by death of the tissue supplied by the vessel whereas gradual occlusion may allow dilation of the anastomosis and adequate nourishment of the tissue. Some arteries have no anastomosis with other arteries and these are called end-arteries. Occlusion of an end-artery will cause death (necrosis) of the tissue supplied by the vessel; an example is the central artery of the retina, following occlusion of which permanent blindness will occur.

Arteries

Arteries are named after the bones in the limbs or from the organ they supply. The most important point about them is their position: where they can be compressed against a bone the *pulse* may be felt or pressure may be applied as a First Aid measure in haemorrhage but at the same time it is important to ensure that there is no undue pressure being exerted on them by splints or other equipment. They are usually found on the inner side of a limb where there is only one bone, or between the bones where there are two, and at a joint they would cross the flexor surface. These positions are the safest because the vessels are less likely to be exposed to injury and pressure.

Aorta

The *aorta* is the main artery of those that carry oxygenated blood to the tissues of the body (Fig. 16.1). It arises from the upper part of the left ventricle, passes upwards and to the right (the *ascending aorta*) and then arches backwards to the left (the arch of the aorta) and passes down through the thorax on the left side of the spine (the *descending thoracic aorta*). It enters the abdominal cavity through an opening in the diaphragm called the *aortic hiatus* and is then called the *abdominal aorta*. It ends at the lower border of the fourth lumbar vertebra by dividing into the right and left common *iliac arteries*.

The ascending aorta has two branches, the right and left *coronary arteries*, which arise immediately above the cusps of the aortic valve. They supply the heart wall.

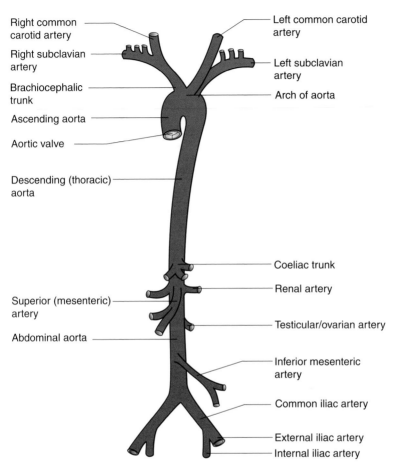

Right common carotid artery

Right subclavian artery

Brachiocephalic trunk

Ascending aorta

Aortic valve

Descending (thoracic) aorta

Superior (mesenteric) artery

Abdominal aorta

Left common carotid artery

Left subclavian artery

Arch of aorta

Coeliac trunk

Renal artery

Testicular/ovarian artery

Inferior mesenteric artery

Common iliac artery

External iliac artery

Internal iliac artery

Fig. 16.1   The arch of the aorta and branches.

The arch of the aorta has three branches, which arise from the top of the arch:

- Brachiocephalic trunk, which divides into the right subclavian artery and the right common carotid artery
- Left common carotid artery
- Left subclavian artery.

The *common carotid arteries* supply the head and neck. Each divides into two at the level of the thyroid cartilage to form the *external* and *internal carotid arteries*. The external carotid artery supplies the outer parts of the face and scalp and has many branches

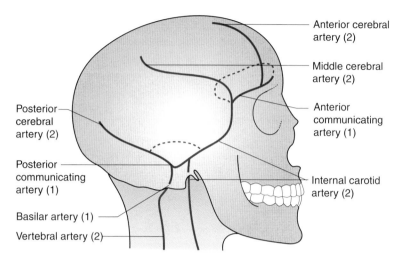

Fig. 16.2 Blood supply to the brain (side view). Broken lines indicate communication with the opposite side of the brain.

such as the *facial, temporal, occipital* and *maxillary branches*. The internal carotid artery supplies a large part of the cerebrum, the eyes and the nose and forehead. At the point where the common carotid artery divides there is a dilated area called the *carotid sinus* where a large number of sensory nerve endings from the glossopharyngeal (ninth cranial) nerve are situated. The sinus reacts to changes in arterial blood pressure and assists in making appropriate changes to return it to normal. A small reddish-brown structure behind the division of the common carotid artery is called the *carotid body*; it acts as a chemoreceptor (see pp 229 and 273).

The vertebral arteries arise from the first part of the *subclavian arteries* and pass upwards in the foramina of the transverse processes of the cervical vertebrae; they enter the skull through the foramen magnum (see Fig. 10.3) and join together to form the *basilar artery*.

An anastomosis named the *circulus arteriosus* connects the vertebral arteries and the two internal carotid arteries. This circle is situated at the base of the brain (Fig. 16.2). In front the two anterior cerebral arteries are joined by the anterior communicating artery (Fig. 16.3). Behind, the basilar artery, formed by the junction of the two vertebral arteries, divides into two posterior cerebral arteries, each of which is joined to the internal carotid arteries by the *posterior communicating artery*. This anastomosis increases the likelihood of

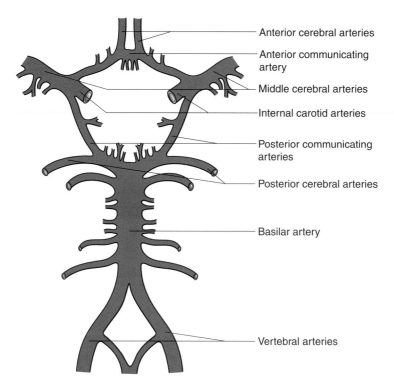

Anterior cerebral arteries

Anterior communicating
artery

Middle cerebral arteries

Internal carotid arteries

Posterior communicating
arteries

Posterior cerebral arteries

Basilar artery

Vertebral arteries

Fig. 16.3   The arterial circle of the cerebrum (circle of Willis).

an adequate blood supply being maintained to the brain if one of
the vessels is injured or occluded.

Blood is supplied to the upper limb through one main artery
called the subclavian artery as far as the outer border of the first
rib, the axillary artery as far as the middle third of the humerus and
the brachial artery as far as the back of the radius, where it divides
into the radial and ulnar arteries. The *radial artery* passes along the
radial side of the forearm to the wrist, where its pulsation can be felt
quite easily, and then crosses into the palm of the hand to form the
deep palmar arch and to unite with the deep branch of the ulnar
artery.

The *ulnar artery* runs down the inner side of the forearm to the
wrist, which it crosses to form the superficial palmar arch. The
palmar arches supply the digits and each anastomoses freely with
the other so that injury to one allows blood to be brought to the part
by other vessels.

The descending thoracic aorta is situated in the mediastinum; it supplies the pericardium, bronchi, oesophagus, mediastinum, intercostal muscles and breasts through branches that are named after the part they supply.

The *abdominal aorta* begins at the aortic hiatus of the diaphragm, about the level of the last thoracic vertebra. It has many large branch arteries arising from it to supply the abdominal organs and therefore it decreases rapidly in size (Fig. 16.4). The branches of the abdominal aorta are:

- Phrenic arteries – supply the diaphragm

- Coeliac trunk – arises just below the aortic opening in the diaphragm and divides into three branches:
    - the left *gastric artery*, which supplies the stomach and gives off two or three branches that ascend through the oesophageal opening in the diaphragm and anastomose with the oesophageal arteries
    - the *hepatic artery*, which supplies the liver and gives off branches to the duodenum and bile duct as well as supplying the right gastric artery
    - the *splenic artery*, which divides into many branches to supply the very vascular spleen and also gives off numerous small vessels to supply the pancreas
- Superior mesenteric artery – supplies all the small intestine, except part of the duodenum, and the first part of the large intestine

- Middle adrenal arteries – arise, one on each side of the aorta, opposite the superior mesenteric artery and supply the adrenal glands

- Renal arteries – supply the kidneys; the left artery is a little higher than the right because of the positions of the respective kidneys

- Ovarian arteries in women and the testicular arteries in men – supply the organs after which they are named.

- Inferior mesenteric artery – supplies the remainder of the large intestine including the sigmoid colon and rectum.

The abdominal aorta divides into two common *iliac arteries* and these again divide, at the level of the last lumbar intervertebral disc, into the *internal iliac artery*, which supplies the pelvis, the perineum and the gluteal region, and the *external iliac artery*, which supplies the lower limb.

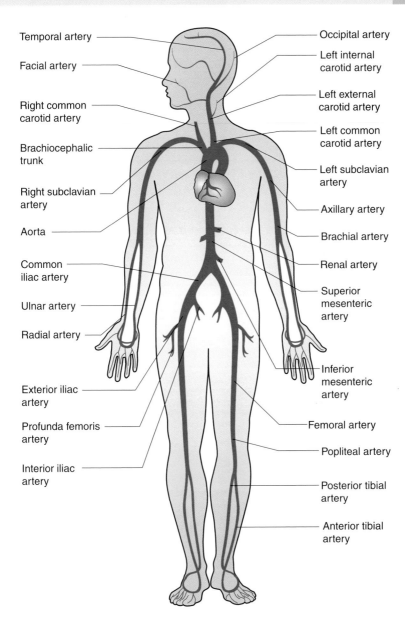

Fig. 16.4   The main arteries.

Blood is supplied to the lower limb through the external iliac artery, which crosses the groin and enters the thigh as the *femoral artery*; this in turn becomes the *popliteal artery* at the lower third of the thigh. The popliteal artery divides into:

1. The *anterior tibial artery*, which runs down the front of the leg on the anterior surface of the interosseous membrane, crosses the front of the ankle joint and supplies the front of the foot as the *dorsalis pedis artery*.

2. The *posterior tibial artery*, which runs down the back of the leg to the back of the ankle joint and to the sole of the foot to become the plantar arch.

The arteries around the ankle joint anastomose freely with one another to form networks of vessels.

Veins

The veins, which return venous blood to the heart, are either *superficial veins*, which vary considerably in position, or *deep veins*, which usually accompany arteries. Arterial pulsation is one of the factors that aids venous return. Systemic veins are more variable than

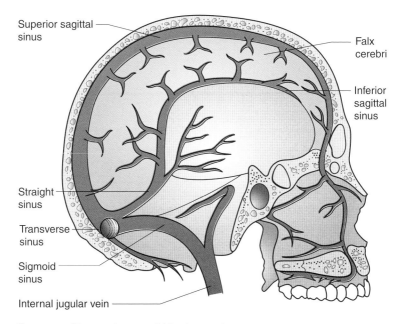

Superior sagittal sinus

Falx cerebri

Inferior sagittal sinus

Straight sinus

Transverse sinus

Sigmoid sinus

Internal jugular vein

Fig. 16.5   Venous sinuses within the cranium.

corresponding arteries and anastomoses occur more frequently. In some areas, such as the pelvis and around the vertebral column, the veins form extensive anastomoses and often do not have valves.

The veins can be divided into two groups:

- Veins of the head, neck, upper limbs and thorax, which all end in the *superior vena cava*
- Veins of the lower limbs, abdomen and pelvis, which all end in the *inferior vena cava*.

The blood from the brain is collected into vessels situated between the two layers of the dura mater and are called *venous sinuses*. These in turn empty into the *internal jugular veins* along with blood from the neck and superficial parts of the face (Fig. 16.5).

The *external jugular veins* receive blood from the exterior of the cranium and from the deep parts of the face. At the root of the neck the internal jugular veins join with the subclavian veins to form the *brachiocephalic veins*, which in turn unite to form the *superior vena cava* through which blood is poured into the right atrium (Fig. 16.6).

The veins of the upper and lower limbs are in two groups: superficial and deep. The superficial veins (Figs 16.7 and 16.8) lie immediately

Right   **Left**

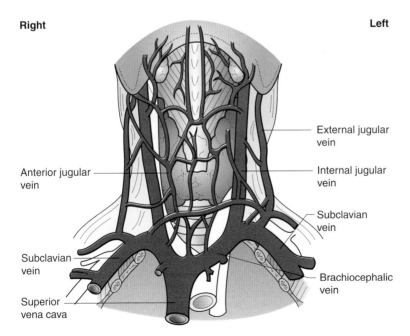

External jugular vein

Anterior jugular vein

Internal jugular vein

Subclavian vein

Subclavian vein

Brachiocephalic vein

Superior vena cava

Fig. 16.6   The chief veins of the neck.

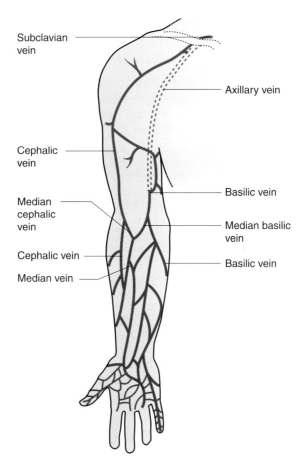

Fig. 16.7 The superficial veins of the forearm.

under the skin; the deep veins follow the course of the arteries (Fig. 16.9). The superficial and deep veins of the upper limb are:

- Superficial veins – the *cephalic, basilic* and *median veins* and their tributaries
- Deep veins – these pour their blood into the axillary vein, which is a continuation of the *basilic veins,* and which becomes continuous with the *subclavian veins.*

While the veins of the lower limbs are:

- Superficial veins – the short and long *saphenous veins,* which empty into the *deep popliteal vein* at the back of the knee

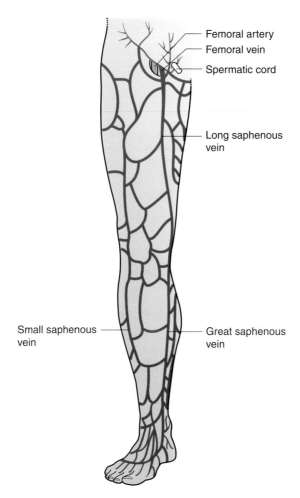

Fig. 16.8 The superficial veins of the lower limb.

- Deep veins – the *anterior and posterior tibial veins*, the *popliteal vein* and the *femoral vein*.

Between the deep and superficial veins of the lower limbs a number of 'perforating' veins exist. These have their valves so arranged that normally blood is prevented from flowing from the deep to the superficial veins. If these valves become ineffective, blood can flow from the deep to the superficial veins, raising the pressure in the latter and causing the dilation and degeneration known as varicose veins and ulcers.

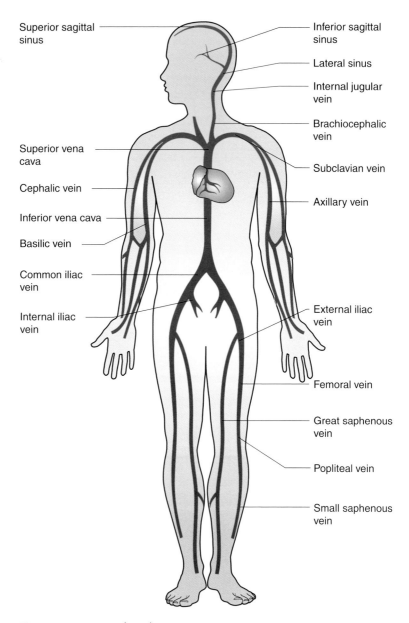

Superior sagittal sinus

Inferior sagittal sinus

Lateral sinus

Internal jugular vein

Brachiocephalic vein

Superior vena cava

Subclavian vein

Cephalic vein

Axillary vein

Inferior vena cava

Basilic vein

Common iliac vein

Internal iliac vein

External iliac vein

Femoral vein

Great saphenous vein

Popliteal vein

Small saphenous vein

Fig. 16.9 The deep (main) veins.

The *femoral vein* ends at the inguinal ligament by becoming the *external iliac vein*. The *internal iliac vein*, returning blood from the pelvis, unites with the external iliac vein to become the *common iliac vein*, one on either side of the body. These unite at the level of the fifth lumbar vertebra to form the *inferior vena cava*, which receives several tributaries, of which some important ones are:

- Renal veins – from the kidneys
- Ovarian or testicular veins – from the reproductive organs, which empty into the inferior vena cava on the right and the renal vein on the left
- Hepatic veins – which return all the blood from the liver, including that from the portal circulation.

The *venae cavae* pour all venous blood into the right atrium except that from the coronary circulation, which is conveyed by the coronary sinus directly into the right atrium.

THE PULMONARY CIRCULATION

These vessels are concerned with carrying deoxygenated blood from the heart to the lungs and oxygenated blood from the lungs back to the heart (see also Chapter 18). The *pulmonary trunk* carries venous blood from the right ventricle; at the left of the fifth thoracic vertebra it divides into the right and left *pulmonary arteries*, which then branch to carry blood to the various segments of the lungs. These vessels are the only arteries carrying deoxygenated blood.

The four pulmonary veins return the oxygenated blood from the lungs to the left atrium, two vessels from each lung. These are the only veins that carry oxygenated blood.

THE PORTAL CIRCULATION

The portal system includes all the veins that drain blood from the abdominal part of the digestive system and from the spleen, pancreas and gall bladder. The blood from these organs is carried to the liver by the *portal vein*, which ends in vessels like capillaries, called sinusoids. The blood is then conveyed by the hepatic veins to the inferior vena cava. As mentioned above the liver must also receive oxygenated blood and this is conveyed to it by the *hepatic artery*.

THE FETAL CIRCULATION

Fetal blood is carried to and from the placenta by *umbilical arteries and veins*. Most of the blood that enters the right atrium through the inferior vena cava passes through an opening in the atrial septum, called the *foramen ovale*, directly into the left atrium and then into the left ventricle and aorta. The blood returning to the right atrium through the superior vena cava passes into the right ventricle and pulmonary trunk, after which only a small part of it goes to the lungs. Most of it passes through the ductus arteriosus directly to the aorta.

At birth the foramen ovale closes so that blood cannot pass from the right atrium to the left atrium but is directed into the pulmonary trunk and then to the lungs. The ductus arteriosus also closes shortly after birth.

SELF-TEST QUESTIONS

- What is a collateral circulation?
- What do arteries carry?
- Describe the portal circulation.

Chapter 17

The lymphatic system

LEARNING OBJECTIVES

After reading this chapter you should understand:

- the structure of the lymphatics
- the function of the lymphatics

As blood passes through the capillaries in the tissues, fluid oozes out through the porous walls and circulates through the tissues themselves, bathing every cell. This fluid is called tissue or *interstitial fluid*; it fills the interstices or the spaces between the cells that form the different tissues (Fig. 17.1). It is a clear, watery, straw-coloured fluid similar to the plasma of the blood from which it is derived. While blood circulates only through the blood vessels, tissue fluid circulates through the actual tissue and carries nutrients, oxygen and water from the blood stream to each individual cell and carries away its waste products such as carbon dioxide, urea and water, transmitting them to the blood. It is, in other words, the carrying medium between the tissue cells and the blood.

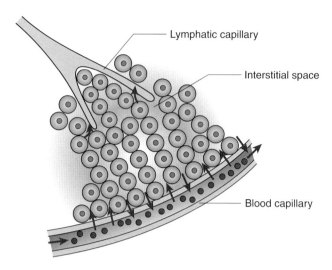

Fig. 17.1 Diagram showing the circulation of tissue fluid. Fluid passes out into the tissues and is collected partly by the blood capillary and partly by the lymph capillary.

Of the fluid that escapes from the capillaries into the tissues, a certain amount passes back through the capillary wall but its return is more difficult than its escape owing to the constant stream of oncoming blood that fills the capillaries. The excess fluid that cannot return directly into the blood stream is collected and returned to the blood by a second set of vessels, which form the lymphatic system; the fluid that these vessels contain is called *lymph*. If fluid accumulates in the interstitial tissues and is not removed by the lymphatic system, oedema results.

THE COMPONENTS OF THE LYMPHATIC SYSTEM

The lymphatic system comprises four types of structure:

- Lymphatic capillaries
- Lymphatic vessels
- Lymphatic nodes
- Lymphatic ducts.

Lymphatic capillaries

The lymphatic capillaries arise in the spaces in the tissues as fine hair-like vessels with porous walls. These gather up excess fluid from the

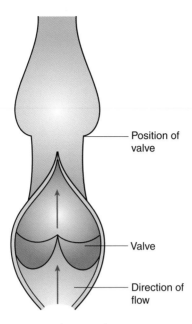

Position of valve

Valve

Direction of flow

Fig. 17.2   A lymphatic vessel (cut open).

tissues and unite to form the lymphatic vessels. The walls of lymph capillaries are permeable to substances of greater molecular size than those that can pass through the walls of blood capillaries.

Lymphatic vessels

The lymphatic vessels are thin-walled, collapsible tubes similar in structure to the veins but carrying lymph instead of blood. They are finer and more numerous than the veins and, like them, are provided with valves to prevent the lymph moving in the wrong direction (Fig. 17.2). Lymphatic vessels are found in most tissues except the central nervous system but run particularly in the subcutaneous tissues and pass through one or more lymphatic nodes (Figs 17.3–17.5).

Lymphatic nodes

The lymphatic nodes are small bodies varying in size from a pinhead to an almond. Lymphatic vessels bring lymph to them and are called afferent vessels. These enter and divide up within the node and discharge the lymph into the lumen. The lymph is then gathered again

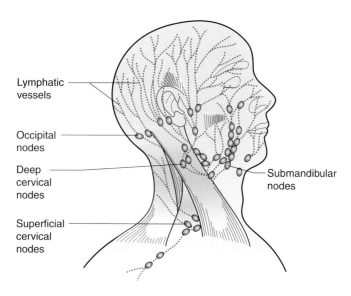

Fig. 17.3   The lymphatic vessels of the head and neck.

into fresh lymphatic vessels, called efferent vessels, which carry it on and ultimately empty it into the lymphatic ducts after possibly carrying the lymph through more nodes (Fig. 17.6). The lymphatic nodes consist essentially of cells similar to the white blood corpuscles (lymphocytes), held together by a network of connective tissue, which also forms a capsule for the node.

The functions of the lymphatic nodes are:

- To filter the lymph of bacteria as it passes through. Thus when tissues are infected the node may become swollen and tender. If the infection is a mild one, the organisms will be overcome by cells of the node, and the tenderness and swelling will subside. If it is severe the organisms will cause acute inflammation and destruction of the white cells may result, causing an abscess to form in the node. If the bacteria are not destroyed by the node, they may pass on in the lymph stream and infect the general circulation, causing septicaemia.

- To provide fresh lymphocytes for the blood stream. The cells of the node constantly multiply and the newly formed cells are carried away in the lymph.

- To produce some antibodies and antitoxins to prevent infection.

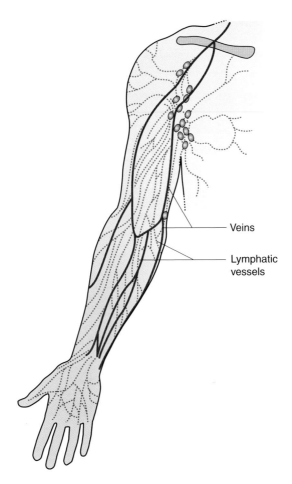

Fig. 17.4   The lymphatic vessels of the upper limb.

The lymphatic nodes are for the most part massed together in groups in various parts of the body. Groups in the neck and under the chin filter the lymph from the head, tongue and the floor of the mouth. A group in the axilla filters the lymph from the upper limb and the chest wall. A group in the groin filters lymph from the lower limb and lower abdominal wall. Groups within the thorax and abdomen filter the lymph from the internal organs.

Special areas where much lymphatic tissue is found include the *palatine* and *pharyngeal tonsils*, the *thymus gland*, aggregated *lymphatic follicles* in the small intestine, the *appendix* and the *spleen*.

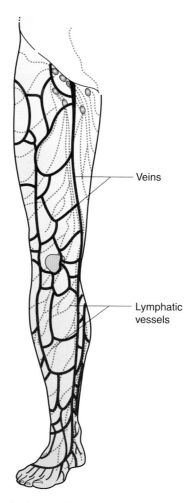

Veins

Lymphatic vessels

Fig. 17.5   The lymphatic vessels of the lower limb.

Lymphatic ducts

After filtration by the nodes the lymph is emptied by the lymphatic vessels into the two lymphatic ducts: the *thoracic duct* and *right lymphatic duct* (Fig. 17.7).

The thoracic duct is the larger. It begins in a small pouch at the back of the abdomen called the *cisterna chyli*. Into this, empty all the lymphatic vessels from the lower limbs and the abdominal and pelvic organs. From the cisterna chyli the duct runs up through the

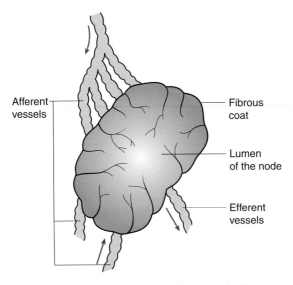

Fig. 17.6   A lymphatic node, showing the afferent and efferent vessels.

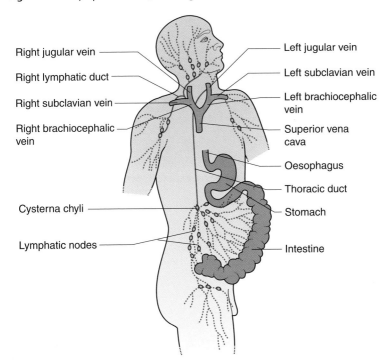

Fig. 17.7   The lymphatic ducts.

mediastinum behind the heart to the root of the neck, where it turns to the left, is joined by the lymphatic vessels from the left side of the head and thorax and the left upper limb, and finally empties into the left subclavian vein at its junction with the left internal jugular vein. It is about 45 cm long and is provided with valves to prevent the lymph flowing in the wrong direction.

The right lymphatic duct is a comparatively small vessel formed by the joining of the lymphatic vessels from the right side of the head and thorax and the right upper limb at the root of the neck. It is only about 1 cm long and enters into the right subclavian vein, where it joins the right internal jugular vein.

The lymphatic ducts thus gather up all the lymph and return it to the blood stream, from which the fluid in the tissues is constantly renewed.

THE FUNCTIONS OF THE LYMPHATIC SYSTEM

The functions of the lymphatic system are:

1. The lymphatic vessels gather up the excessive fluid or lymph from the tissues and thus permit a constant stream of fresh fluid to circulate through them.
2. It is the channel by which excess proteins in the tissue fluid pass back into the blood stream.
3. The nodes filter the lymph of bacterial infection and harmful substances.
4. The nodes produce fresh lymphocytes for the circulation.
5. The lymphatic vessels in the abdominal organs assist in the absorption of digested food, especially fat.

THE MECHANISM OF THE LYMPHATIC CIRCULATION

The lymphatic circulation is maintained partly by 'suction', partly by pressure. Suction is the more important factor. The lymphatics empty into the large veins approaching the heart, and here there is negative pressure due to suction as the heart expands, and also suction towards the thorax during the act of inspiration.

Pressure is exerted on the lymphatics, as on the veins, by the contraction of muscles, and this outside pressure drives lymph onwards because the valves prevent a backward flow. There is also a slight pressure from the fluid in the tissues, on account of the constant pouring out of fresh fluid from the capillaries. If there is obstruction to the flow of lymph through the lymphatic system, lymphoedema results, i.e. a swelling of the tissues due to excess fluid collecting in them. The condition also results from any obstruction of the veins, since these also drain fluid from the tissues. Lymphoedema is quite common in some cancers where the lymphatics are involved, such as breast cancer, where lymphoedema of the arms occurs.

THE SPLEEN

The spleen is a large nodule of lymphoid tissue (Fig. 17.8). By function it belongs to the circulatory system, as do the lymphatic nodes. It is deep purplish-red and lies high up at the back of the abdomen on the left side behind the stomach. It is enclosed in a capsule of fibrous tissue, and fibrous strands make a supporting meshwork throughout the gland. The spaces of this meshwork are filled by a pulp-like material, called the *splenic pulp*, which is the essential substance of the organ and contains cells of different types. Many of these are similar to the lymphocytes of the blood and lymph nodes, and these help to produce fresh white cells for the blood stream. Others are phagocytes (devouring cells), which engulf the red blood cells that are beginning to wear out and break them down.

The functions of the spleen are not fully known but the spleen is thought to be:

1. A source of fresh lymphocytes for the blood stream.
2. A seat for the destruction of red blood cells.

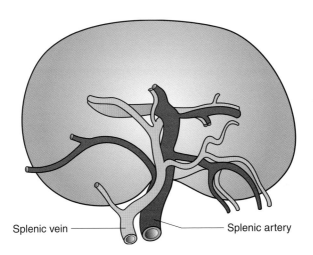

Splenic vein ————————————————— Splenic artery

Fig. 17.8 The spleen and its blood vessels.

The spleen is also thought to assist in fighting infection as it becomes enlarged in certain diseases where the blood is infected, e.g. malaria and typhoid fever. It probably helps in the manufacture of antibodies to fight infection. However, it is not essential to life and can be removed when it is responsible for ill health, e.g. in haemolytic anaemia.

SELF-TEST QUESTIONS

- Where are the lymph nodes located?
- How is lymph circulated?

Chapter **18**

The respiratory system

CHAPTER CONTENTS

LEARNING OBJECTIVES

After reading this chapter you should understand:

- the structures of the respiratory system
- the structure of the lungs
- the mechanism of ventilation of the lungs
- the function of respiration
- the control of respiration
- the lung volumes

All living cells require a constant supply of oxygen in order to carry on their metabolism. Oxygen is in the air, and the respiratory system is constructed in such a way that air can be taken into the lungs, where some of the oxygen is extracted for use by the body and at the

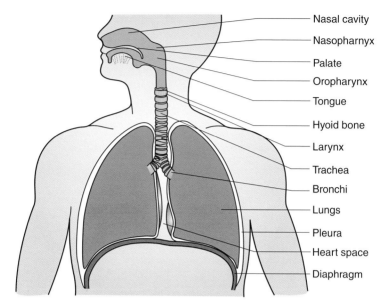

Fig. 18.1 Diagram of the respiratory tract.

same time carbon dioxide and water vapour are given up. The organs of the respiratory system (Fig. 18.1) are:

- The nose
- The pharynx
- The larynx   } leading to the lungs
- The trachea
- The bronchi
- The bronchioles } within the lungs.
- The alveolar ducts and alveoli.

THE NOSE

The external nose is the visible part of the nose, formed by the two nasal bones and by cartilage. It is both covered and lined by skin and inside there are hairs that help to prevent foreign material from entering. The *nasal cavity* (Fig. 18.2) is a large cavity divided by a *septum*. The *anterior nares* are the openings that lead in from the outside world and the *posterior nares* are similar openings at the back, leading in to the pharynx. The roof is formed by the ethmoid bone at the base of the skull and the floor by the hard and soft palates at the roof

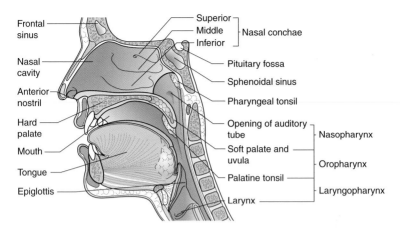

Fig. 18.2 Sagittal section of the nose, mouth, pharynx and larynx.

of the mouth. The lateral walls of the cavity are formed by the maxilla, the superior and middle nasal conchae of the ethmoid bone, and the inferior nasal concha. The posterior part of the dividing septum is formed by the perpendicular plate of the ethmoid bone and by the vomer, while the anterior part is made of cartilage. The bones of the cranium and the face have already been described on pages 132–138.

The three *nasal conchae* project into the nasal cavity on each side and greatly increase the surface area of the inside of the nose. The cavity of the nose is lined throughout with *ciliated mucous membrane*, which is extremely vascular; atmospheric air is warmed as it passes over the epithelium, which contains many capillaries. The mucus moistens the air and entraps some of the dust, and the cilia move the mucus back into the pharynx for swallowing or expectoration. The nerve endings of the sense of smell are situated in the highest part of the nasal cavity round the cribriform plate of the ethmoid bone.

Some of the bones surrounding the nasal cavity are hollow. The hollows in the bones are called the *paranasal sinuses*, which both lighten the bones and act as sounding chambers for the voice, making it resonant. The *maxillary sinus* lies below the orbit and opens through the lateral wall of the nose. The *frontal sinus* lies above the orbit towards the midline of the frontal bone. The *ethmoidal sinuses* are numerous and are contained within the part of the ethmoid bone separating the orbit from the nose, and the sphenoidal sinus is in the body of the sphenoid bone. All the paranasal sinuses are lined with mucous membrane and all open into the nasal cavity, from which they may become infected.

THE PHARYNX

The roof of the pharynx is formed by the body of the sphenoid bone and, inferiorly, it is continuous with the oesophagus (see Fig. 18.2). At the back it is separated from the cervical vertebrae by loose connective tissue, while the front wall is incomplete and communicates with the nose, mouth and larynx. The *pharynx* is divided into three sections: the *nasopharynx*, which lies behind the nose; the *oropharynx*, which lies behind the mouth; and the *laryngopharynx*, which lies behind the larynx.

The nasopharynx is that part of the pharynx that lies behind the nose above the level of the soft palate. On the posterior wall, there are patches of lymphoid tissue called the pharyngeal tonsils, commonly referred to as the *adenoids*. This tissue sometimes enlarges and blocks the pharynx, causing mouth breathing in children. The auditory (or acoustic) tubes open from the lateral walls of the nasopharynx and through them air is carried to the middle ear. The nasopharynx is lined with ciliated mucous membrane, which is continuous with the lining of the nose.

The oropharynx lies behind the mouth below the level of the soft palate, with which its lateral walls are continuous. Between the folds of these walls, which are called the palatoglossal arches, are collections of lymphoid tissue called the *palatine tonsils*. The oropharynx is part of the respiratory tract and alimentary tract but it cannot be used for swallowing and breathing simultaneously. During swallowing, breathing stops momentarily and the oropharynx is completely blocked off from the nasopharynx by the raising of the soft palate. The oropharynx is lined with stratified epithelium.

THE LARYNX

The larynx is continuous with the oropharynx above and with the trachea below (see Fig. 18.2). Above it lies the hyoid bone and the root of the tongue. The muscles of the neck lie in front of the larynx, and behind the larynx lies the laryngopharynx and the cervical vertebrae. On either side are the lobes of the thyroid gland. The larynx is composed of several irregular cartilages joined together by ligaments and membranes (Fig. 18.3).

The thyroid cartilage is formed of two flat pieces of cartilage fused together in the front to form the laryngeal prominence or *Adam's apple*. Above the prominence is a notch called the *thyroid notch*. The thyroid cartilage is larger in the male than in the female. The upper

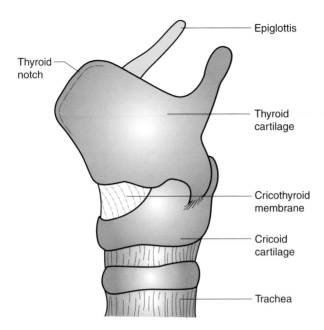

Fig. 18.3   The laryngeal cartilages.

part is lined with stratified epithelium, the lower part with ciliated epithelium.

The *cricoid cartilage* lies below the thyroid cartilage and is shaped like a signet ring with the broad portion at the back. It forms the lateral and posterior walls of the larynx and is lined with ciliated epithelium.

The *epiglottis* is a leaf-shaped cartilage attached to the inside of the front wall of the thyroid cartilage immediately below the thyroid notch. During swallowing the larynx moves upwards and forwards so that its opening is occluded by the epiglottis.

The *arytenoid cartilages* are a pair of small pyramids made of hyaline cartilage. They are situated on top of the broad part of the cricoid cartilage, and the vocal ligaments are attached to them. They form the posterior wall of the larynx.

The hyoid bone and the laryngeal cartilages are joined together by ligaments and membranes. One of these, the *cricothyroid membrane,* is attached all round to the upper edge of the cricoid cartilage and has a free upper border, which is not circular like the lower border but makes two parallel lines running from front to back. The two parallel edges are the vocal ligaments (Fig. 18.4). They are fixed to the middle of the thyroid cartilage in front and to the arytenoid cartilages

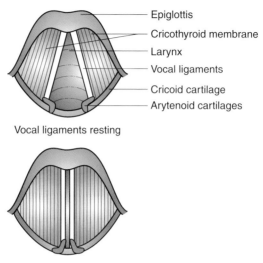

Fig. 18.4 Diagram of the vocal ligaments.

behind, and they contain much elastic tissue. When the intrinsic muscles of the larynx alter the position of the arytenoid cartilage the vocal ligaments are pulled together, narrowing the gap between them. If air is forced through the narrow gap, called the chink, during expiration, the vocal ligaments vibrate and sound is produced. The pitch of the sound produced depends on the length and tightness of the ligaments: an increased tension gives a higher note; a slacker tension a lower note. Loudness depends on the force with which the air is expired. The alteration of the sound into different words depends on the movements of the mouth, tongue, lips and facial muscles. An inflammation of the larynx is called laryngitis and this may be caused by infections or chemicals. Laryngitis interferes with speech and leads to a hoarse voice.

THE TRACHEA

The trachea begins below the larynx and runs down the front of the neck into the chest. It divides into the right and the left main *bronchi* at the level of the fifth thoracic vertebra and is about 12 cm long. The isthmus of the thyroid gland crosses in front of the upper part of the trachea, and the arch of the aorta lies in front of the lower part, with the manubrium of the sternum in front of it. The oesophagus lies behind the trachea, separating it from the bodies of the thoracic verte-brae. On either side of the trachea lie the lungs, with the lobes of the thyroid gland above them. The wall of the trachea is made of invol-untary muscle and fibrous tissue strengthened by incomplete rings of hyaline cartilage. The deficiency in the cartilage lies at the back where the trachea is in contact with the oesophagus. When a bolus of food is swallowed the oesophagus is able to expand without hin-drance but the cartilage maintains the patency of the airway. The trachea is lined with ciliated epithelium containing *goblet cells*, which secrete mucus. The cilia sweep the mucus and foreign particles upwards towards the larynx.

THE LUNGS

The lungs are two large spongy organs lying in the thorax on either side of the heart and great vessels. They extend from the root of the neck to the diaphragm and are roughly conical, with the apex above and the base below. The ribs, costal cartilages and intercostal muscles lie in front of the lungs, and behind them are the ribs, the intercostal muscles and the transverse processes of the thoracic vertebrae. Between the lungs is the mediastinum, which completely separates one side of the thoracic cavity from the other, stretching from the vertebrae behind to the sternum in front. Within the mediastinum lie the heart and great vessels, the trachea and the oesophagus, the thoracic duct and the thymus gland.

The lungs are divided into *lobes* (Fig. 18.5). The left lung has two lobes, separated by the oblique fissure. The upper lobe is above and in front of the lower lobe, which is conical. The right lung has three lobes. The lower lobe is separated by an oblique fissure in a similar manner to the left lower lobe. The remainder of the lung is separated by a hori-zontal fissure into the upper and middle lobes. Each lobe is further divided into named *bronchopulmonary segments*, separated from each other by a wall of connective tissue and each having an artery and a vein. Each segment is also divided into smaller units called *lobules*.

The bronchi

The two main bronchi commence at the bifurcation of the trachea and one leads into each lung. The left main bronchus is narrower, longer and more horizontal than the right main bronchus as the heart lies a little to the left of the midline. Each main bronchus divides into branches, one for each lobe. Each of these then divides

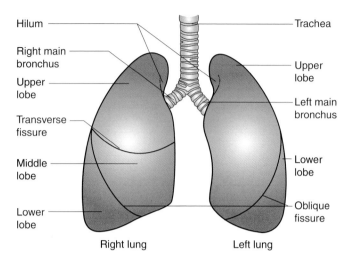

Fig. 18.5 Diagram showing the lobes of the lungs.

into named branches, one for each bronchopulmonary segment, and divides again into progressively smaller bronchi within the lung substance. The bronchi are similar in structure to the trachea but the cartilage is less regular. Inflammation of the bronchi is called bronchitis.

The bronchioles

The finest bronchi are called *bronchioles*. They have no cartilage but are composed of muscular, fibrous and elastic tissue lined with cuboid epithelium. As the bronchioles become smaller, the muscular and fibrous tissue disappears and the smallest tubes, called terminal bronchioles, are a single layer of flattened epithelial cells. The bronchioles are affected in asthma, which leads to their constriction and creates difficulty with breathing.

The alveolar ducts and alveoli

The *terminal bronchioles* branch repeatedly to form minute passages called *alveolar ducts*, from which *alveolar sacs* and alveoli open. The alveoli are surrounded by a network of capillaries (Fig. 18.6). Deoxygenated blood enters the capillary network from the pulmonary artery and oxygenated blood leaves it to enter the pulmonary veins.

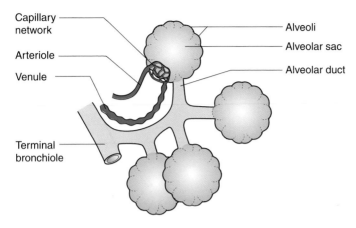

Fig. 18.6   Diagram of alveoli.

It is in the capillary network that the exchange of gases takes place between the air in the alveoli and the blood in the vessels.

The hilum of the lung

The *hilum* is a triangular depression on the concave medial surface of the lung. The structures forming the root of the lung enter and leave at the hilum, which is at the level of the fifth to seventh thoracic vertebrae. These structures include the main bronchus, the pulmonary artery, the bronchial artery and branches of the vagus nerve, which enter at this point, and two pulmonary veins, the bronchial veins and lymphatic vessels, which leave the lung at the root. There are also many lymph nodes around the root of the lung.

The pleura

The *pleura* is a serous membrane, which surrounds each lung. It is composed of flattened epithelial cells on a basement membrane and it has two layers (Fig. 18.7). The *visceral pleura* is firmly attached to the lungs, covering their surfaces and dipping into the interlobar fissures. At the root of the lungs the visceral layer is reflected back to become the *parietal layer*, which lines the chest wall and covers the superior surface of the diaphragm. The two layers of the pleura are normally in close contact with each other, separated only by a film of serous fluid that enables them to glide over one another without friction. This *potential space* between the layers is called the *pleural cavity*.

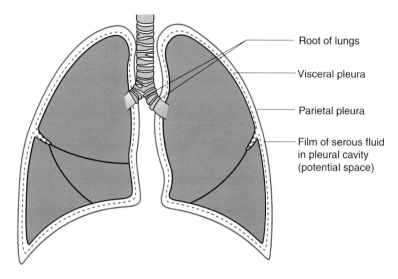

Root of lungs

Visceral pleura

Parietal pleura

Film of serous fluid in pleural cavity (potential space)

Fig. 18.7 Diagram of the pleura.

The pressure between the pleura is lower than atmospheric pressure, i.e. it is negative, and this is what holds the two pleural membranes together. If the pleura become punctured from either the outside of the body, by injury, or spontaneously from within the lungs, then air fills the potential space and the lungs collapse – a condition known as pneumothorax.

MECHANISM OF RESPIRATION

Respiration consists of two phases, inspiration and expiration. The chest expands during inspiration owing to movement of the diaphragm and intercostal muscles. When the diaphragm contracts during inspiration it is flattened and lowered and the thoracic cavity is increased in length (see also Chapter 13). The external intercostal muscles, on contraction, lift the ribs and draw them out, increasing the depth of the thoracic cavity. As the chest wall moves up and out the parietal pleura, which is closely attached to it, moves with it. The visceral pleura follows the parietal pleura and the volume of the interior of the thorax is increased. The lung expands to fill the space and air is sucked into the bronchial tree.

Expiration during quiet breathing is passive. The diaphragm relaxes and assumes its original domed shape. The intercostal muscles relax and the ribs revert to their previous position. The lungs recoil

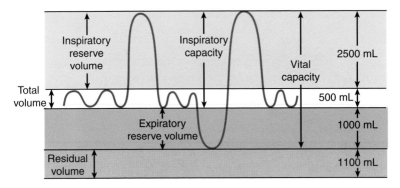

Fig. 18.8   Lung volumes and capacities.

and air is driven out through the bronchial tree. In forced expiration the internal intercostal muscles contract actively to lower the ribs. The *accessory muscles of respiration* may be brought into use during deep breathing or when the airway is obstructed. During inspiration the sternocleidomastoid muscles raise the sternum and increase the diameter of the thorax from front to back. Serratus anterior and pectoralis major pull the ribs outwards when the arm is fixed. Latissimus dorsi and the muscles of the anterior abdominal wall help to compress the thorax during forced expiration.

Control of respiration

Respiration is controlled by the *respiratory centre* in the medulla oblongata. The presence of an accumulation of carbon dioxide in the blood stimulates chemoreceptors in the great arteries. Impulses are carried by the *vagus* and *glossopharyngeal nerves* to the respiratory centre, from the respiratory centre by the *phrenic nerves* to the diaphragm and by the intercostal nerves to the intercostal muscles. These impulses cause the muscles to contract and inspiration occurs.

THE CAPACITY OF THE LUNGS

Tidal volume is the amount of air breathed in and out during normal quiet breathing (about 500 mL). After a normal expiration the amount of air taken in during forced inspiration is called the *inspiratory capacity* (about 3000 mL). It includes the tidal volume. After a quiet expiration, it is possible to force another 1000 mL of air from the lungs. This is the *expiratory* reserve volume (Fig. 18.8).

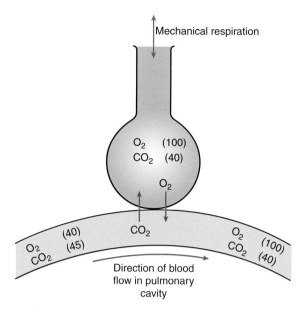

Fig. 18.9 Gas exchange in the lungs. (All values in parentheses are in units of mmHg.)

After the deepest possible expiration, there is still the *residual volume* (about 1100 mL) left in the respiratory passages.

Vital capacity is the largest volume of air expired after the deepest possible inspiration (about 4000 mL). An important measure of respiratory function is the amount of forced expiration in one second. It is called the *forced expiratory volume* and should measure 75–80% of the vital capacity.

GAS EXCHANGE

Exchange of gases within the body takes place both in the lungs, where it is called *external respiration*, and in the tissues, where it is called *internal respiration*. An elementary law of physics relating to gases states that they tend to diffuse from a higher pressure to a lower pressure (Fig. 18.9). The air breathed in contains several gases. The composition of inspired air is:

- Nitrogen 79%
- Oxygen 21%
- Carbon dioxide 0.04%
- Water vapour
- Traces of other gases.

External respiration

When inspired air reaches the alveoli it is in close contact with the blood in the surrounding capillary network. Oxygen in the alveoli, at a pressure of 100 mmHg, comes into contact with oxygen in the venous blood at a pressure of 40 mmHg and therefore the gas diffuses into the blood until the pressures are equal. At the same time, carbon dioxide in the blood, at a pressure of 46 mmHg, comes into contact with alveolar carbon dioxide at a pressure of 40 mmHg and therefore the gas diffuses out of the blood into the alveoli. The gaseous content of expired air is therefore altered to contain less oxygen and more carbon dioxide. The nitrogen content remains the same. The composition of expired air is:

- Nitrogen 79%
- Oxygen 16%
- Carbon dioxide 4.5%
- Water vapour
- Traces of other gases.

Any condition of the lungs that increases the distance for the diffusion of gases between the alveoli and the lungs decreases the exchange of oxygen and carbon dioxide. Pulmonary oedema and pneumonia, which both lead to the accumulation of fluid in the lungs, are both associated with reduced gas exchange in the lungs.

Internal respiration

The oxygen that has diffused into the blood is carried in the haemoglobin (now called oxyhaemoglobin) to the tissues. Here the pressure of the oxygen is low, so the gas diffuses out of the blood and into the tissues, the amount depending on the activity in the tissue. At the same time carbon dioxide, produced in the tissues, is carried away by the blood.

Oxygen–haemoglobin dissociation

The ability of haemoglobin to bind oxygen at the alveoli and then to release it into the peripheral tissues in a way that is physiologically useful is demonstrated by the oxygen–haemoglobin dissociation curve (Fig. 18.10). This curve is obtained by measuring the amount of haemoglobin that has bound oxygen (the saturation with oxygen) as the amount of oxygen (the partial pressure of oxygen) in the blood is increased. The curve has a characteristic sigmoidal (S) shape. If the curve were hyperbolic it would mean that oxygen was

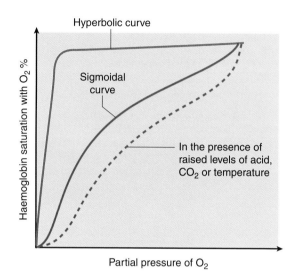

Fig. 18.10 Oxygen–haemoglobin dissociation curve.

not released in the peripheral tissues until the amounts of oxygen had fallen to physiologically dangerous levels. The sigmoidal shape means that oxygen is readily released into peripheral tissues from the haemoglobin in response to falling oxygen levels in order to maintain levels of oxygen that allow these tissues to function. The curve is sensitive to conditions in the peripheral tissues; it moves to the right in response to increased levels of acidity, carbon dioxide or temperature, all of which indicate that there is increased metabolic activity and, therefore, a greater demand for oxygen. The curve moves to the left in fetal blood to make sure that it more readily accumulates oxygen than the maternal blood with which it exchanges oxygen and carbon dioxide in the placenta.

SELF-TEST QUESTIONS

- Explain the function of the intrapleural space.
- Describe gas exchange in the lungs.
- What are tidal volume, vital capacity and residual volume?
- Describe the oxygen–haemoglobin dissociation curve.

SECTION 5

Nutrition and elimination

SECTION INTRODUCTION

Most of this section is concerned with the digestive tract and its associated organs. The digestive tract runs from the mouth to the rectum and is responsible for the ingestion, mixing and mechanical breakdown and absorption of nutrients. Ultimately, those contents of our food that are not absorbed are expelled at the rectum.

The organs associated with digestion are the liver, the pancreas and the gall bladder and these produce and store secretions which are released into the digestive tract and which facilitate the mechanical and chemical breakdown of food. The latter process is called 'digestion' and this prepares food for absorption and subsequent involvement in metabolism. The main features of metabolism are considered.

While it is not anatomically linked to the digestive system, the main function of the urinary system is relevant here as it is responsible for ridding the body of excess toxic substances. Principally these are expelled in urine in the form of urea. The urinary system also responds to changing levels of fluid in the body and can, thereby, conserve water or expel excess water.

Chapter 19

The digestive system

CHAPTER CONTENTS

LEARNING OBJECTIVES

After reading this chapter you should understand:

- the components of the digestive system
- the structure and function of the mouth
- the structure and function of the stomach
- the structure and function of the small intestine
- the process of digestion
- the structure and function of the large intestine
- the structure and function of the peritoneum

The digestive system (Fig. 19.1) consists of all the organs that are concerned in the chewing, swallowing, digestion and absorption of food and in the elimination from the body of indigestible and undigested food. It consists of the digestive tube or *alimentary canal* and the *accessory organs of digestion.*

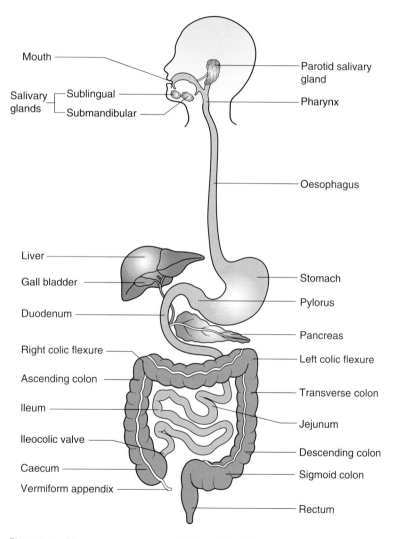

Fig. 19.1 Diagrammatic representation of the digestive system.

The digestive tract is about 9 m long and consists of six parts:

- The mouth
- The pharynx
- The oesophagus
- The stomach
- The small intestine
- The large intestine, which reaches the surface of the body at the anus.

The accessory organs of digestion are:

- The teeth
- Three pairs of salivary glands
- The liver and bile ducts (see Chapter 20)
- The pancreas (see Chapter 20).

THE MOUTH

The mouth is a cavity bounded externally by the lips and cheeks and leading into the pharynx. The roof is formed by the hard and soft *palates*, and the anterior two-thirds of the *tongue* fills the floor of the mouth (Fig. 19.2). The walls are formed by the muscles of the cheeks. The mucous membrane, which lines the mouth, is continuous with the skin of the lips and with the mucous lining of the pharynx. The lips enclose the orbicularis oris muscle, which keeps the mouth closed. The hard palate is formed by parts of the palatine bones and the maxillae; its upper surface forms the floor of the nasal cavity. The soft palate is suspended from the posterior border of the hard palate and extends downwards between the oral and nasal parts of the pharynx. Its lower border hangs like a curtain between the mouth and the pharynx and a small conical process, called the *uvula*, hangs down from it. Two curved folds of mucous membrane extend sideways and downwards from each side of the base of the uvula, called the palatoglossal and palatopharyngeal arches, between which lie the masses of lymphoid tissue known as the palatine tonsils.

Tongue

The tongue is a muscular organ that is attached to the hyoid bone and the mandible. It is covered in certain areas with modifications of the mucous membrane, which appear as projections called *papillae*, to increase the surface area. In addition, specialized areas called *taste buds* (Fig. 19.3) are widespread over almost the entire area of

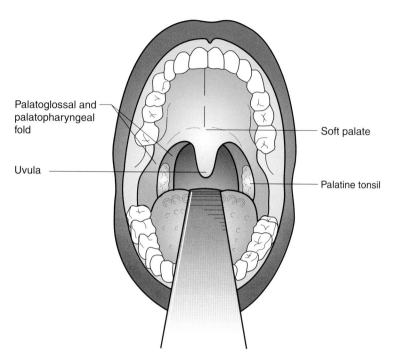

Palatoglossal and palatopharyngeal fold

Uvula

Soft palate

Palatine tonsil

Fig. 19.2   Looking into the mouth with the tongue depressed.

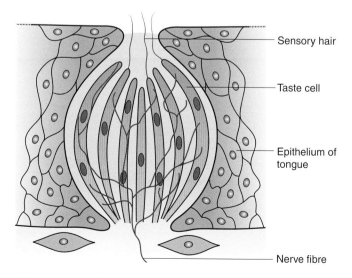

Sensory hair

Taste cell

Epithelium of tongue

Nerve fibre

Fig. 19.3   A taste bud in the tongue.

the tongue. The undersurface of the anterior part of the tongue is connected to the floor of the mouth by a fold of mucous membrane called the *frenulum*. The functions of the tongue are:

1. It is the organ of taste.
2. It assists in the mastication of food.
3. It assists in swallowing.
4. It assists with speech.

Teeth

Humans have two sets of teeth, which make their appearances at different periods of life. The first set are *deciduous* or primary teeth and erupt through the gums during a child's first and second years. The second set begin to replace the first at about the sixth year and the process is usually complete by the twenty-fifth year. Since they cannot be replaced, and may be retained until old age, they are known as the *permanent teeth*.

Each tooth consists of three parts (Fig. 19.4):

- Crown – projecting beyond the gum
- Root – embedded in the alveolus of the maxilla or mandible (a tooth may have one, two or three roots)
- Neck – the constricted part between the crown and the root.

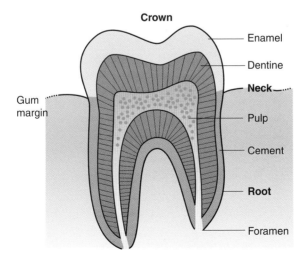

Fig. 19.4   Section of a tooth, showing its structure.

In the centre of all these parts is the *pulp* and immediately outside the pulp is a yellowish-white layer, *dentine*, which forms the main part of the tooth. The outer layer of the tooth is in two parts: that covering the crown is called *enamel* and is a hard, white layer; that covering the root is called *cement* and is a thin layer resembling bone in structure. The pulp is richly supplied with blood vessels and nerves, which enter the tooth through foramina at the apex of each root.

There are four types of teeth:

- Incisor teeth have chisel-shaped crowns giving a sharp cutting edge for biting food
- Canine teeth have large conical crowns
- Premolar (or bicuspid) teeth have almost circular crowns with two cusps for grinding food
- Molar teeth are the largest, and they have broad crowns with four or five cusps.

There are 20 deciduous teeth and 32 permanent teeth (Table 19.1).

Salivary glands

There are three pairs of salivary glands. The *parotid gland* is the largest and lies just below the ear; its duct is about 5 cm long and opens into the mouth opposite the second upper molar tooth. It is this gland that is affected by the disease commonly known as *mumps*. The *submandibular gland* and the *sublingual gland* both open into the floor of the mouth. *Saliva* is secreted by reflex by the presence

Table 19.1   The deciduous teeth and permanent teeth

	The deciduous teeth					
	Molars	Canines	Incisors		Canines	Molars
Upper jaw	2	1	2	2	1	2
Lower jaw	2	1	2	2	1	2

	The permanent teeth							
	Molars	Premolars	Canines	Incisors		Canines	Premolars	Molars
Upper jaw	3	2	1	2	2	1	2	3
Lower jaw	3	2	1	2	2	1	2	3

of food in the mouth and by a conditioned (or learned) reflex that enables saliva to be secreted by the sight, smell or thought of food. Saliva contains a large amount of water, which moistens and softens the food; mucus, which combines the food and lubricates it for its passage down the oesophagus; and the enzyme *salivary amylase*, which acts on cooked starch (carbohydrate) and splits it into maltose and dextrin. Saliva also cleanses the mouth and teeth and keeps the soft parts supple.

THE PHARYNX

When food is well chewed and moistened the tongue rolls it into a *bolus* and carries it towards the oropharynx. The soft palate rises up to occlude the nasopharynx and the epiglottis moves upwards and forwards so that the bolus passes over the closed inlet of the larynx and on into the laryngopharynx and then to the oesophagus. This is an excellent example of muscular co-ordination and if it is not achieved correctly, 'choking' will result (see Fig. 19.2).

THE OESOPHAGUS

The oesophagus is a muscular canal about 25 cm long, extending from the pharynx to the stomach. It begins at the level of the sixth cervical vertebra and descends through the mediastinum in front of the vertebral column and behind the trachea. It passes through the diaphragm at the level of the tenth thoracic vertebra and ends at the cardiac orifice of the stomach at the level of the eleventh thoracic vertebra. On each side of the upper part of the oesophagus are the corresponding common carotid artery and part of the thyroid gland.

The oesophagus has four layers and is similar in structure to the remainder of the alimentary canal (Fig. 19.5):

- Fibrous outer layer – consists of areolar tissue containing many elastic fibres
- Muscular layer – comprised of two layers, the outer layer with fibres running longitudinally and the inner layer consisting of circular fibres
- Areolar or submucous layer – connects the mucous and muscular coats and contains the larger blood vessels and nerves, as well as the mucous glands
- Inner lining of mucous membrane secretes mucus.

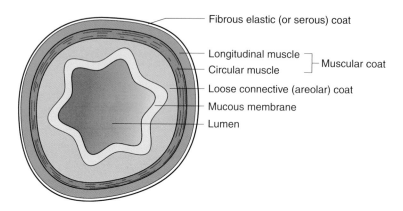

Fig. 19.5   Cross-section of the alimentary tract.

The muscular coat of the upper two-thirds of the oesophagus is of striped voluntary muscle; the lower third contains unstriped involuntary muscle. The oesophagus is innervated by the vagus nerve. Movement of food through the oesophagus is by peristaltic action (*peristalsis* is a wave of dilatation followed by a wave of contraction as succeeding muscle fibres relax and contract.) It takes about 9 seconds for a wave of peristalsis to pass the bolus of food from the pharynx to the stomach.

THE STOMACH

The stomach is the most dilated part of the digestive tube and is situated between the end of the oesophagus and the beginning of the small intestine. Its shape and position are altered by changes within the abdominal cavity and by the stomach contents. It lies below the diaphragm, slightly to the left of the midline (Fig. 19.6).

The stomach is approximately J shaped and has two curvatures. The *lesser curvature* forms the right (or posterior) border of the stomach. The *greater curvature* is directed mainly forwards and first forms an arch upwards and to the left to form the *fundus* of the stomach; it then passes downwards and finally turns right to the point where it joins the *duodenum*. The capacity of the stomach is about 1500 mL in the adult.

The upper opening from the oesophagus is called the *cardiac orifice* and the circular muscle fibres of the oesophagus are slightly thicker at this point and constitute a weak sphincter muscle.

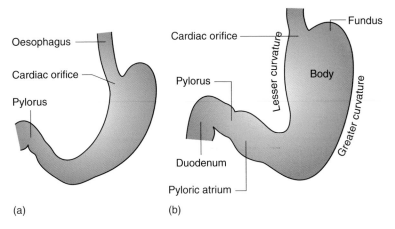

Fig. 19.6 The stomach (a) empty and (b) containing food.

The lower opening, into the duodenum, is called the *pyloric orifice* and is guarded by the strong *pyloric sphincter*, which prevents regurgitation of food from the duodenum into the stomach.

The wall of the stomach consists of four layers (Fig. 19.7):

- The outer serous layer – the visceral layer of the peritoneum (see pp. 300–301)
- The muscular layer – this consists of three layers of smooth muscle fibres, the outer being *longitudinal*, the middle *circular* and the inner *oblique*
- The submucous layer – this is composed of loose areolar tissue
- The lining of mucous membrane – this has a honeycomb appearance because of the presence of the *gastric glands* and their openings. The mucous membrane has numerous folds, called rugae, which run longitudinally and which flatten out when the stomach is full. The mucus secreted by the goblet cells helps to lubricate the food.

The oesophagus penetrates the respiratory diaphragm. If the fundus of the stomach protrudes up through the diaphragm this is referred to as a hiatus hernia and can lead to reflux of the stomach contents up into the oesophagus and therefore symptoms of 'heartburn'.

Functions of the stomach

The functions of the stomach are:

- To churn up the food, breaking it up still further and mixing it with the secretions from the gastric glands. The part of the

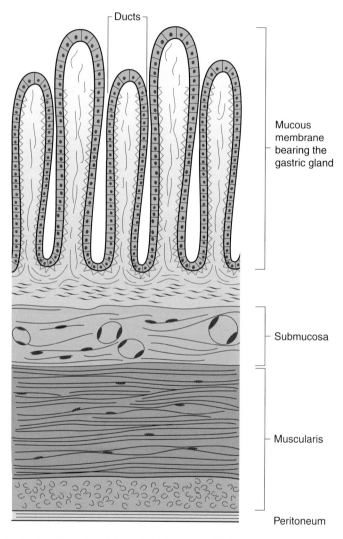

Ducts

Mucous
membrane
bearing the
gastric gland

Submucosa

Muscularis

Peritoneum

Fig. 19.7 Section of gastric wall, highly magnified to show glands.

stomach just before the pyloric sphincter – the *pyloric antrum* – plays a large part in this movement, muscular contraction and relaxation sending some of the liquid food through the sphincter into the small intestine and returning some to the body of the stomach for further mixing

- To continue the digestion of food by means of the *gastric juice*
- To secrete the intrinsic factor (see p. 212).

There are three types of cell in the mucosa of the stomach. *Mucous cells* secrete mucus, which protects the mucous membrane from the action of the other gastric juices. *Chief cells* secrete an enzyme known as pepsinogen (and in children another called rennin) and *oxyntic cells* secrete hydrochloric acid. The secretion of the gastric juice occurs by reflex in the same way as the saliva, causing a copious flow of the fluid before and during the taking of food. The gastric glands are also stimulated by an internal secretion or hormone, produced by the stomach, called *gastrin*, which passes into the circulation and when it reaches the gastric glands increases the production of gastric juice.

Gastric juice

The constituents of gastric juice are:

- Water, mineral salts and mucus
- Hydrochloric acid (HCl)
- Pepsinogen, which is converted by hydrochloric acid into the active enzyme pepsin. Pepsin turns proteins into peptides
- Rennin, which coagulates milk protein into casein, in which form it can be acted upon by pepsin. This is active in babies.

The juice makes the food more liquid and acidic. Until it becomes acidic, and this may take 15–30 minutes in the cardiac end of the stomach which acts as a reservoir, the amylase of the saliva continues to act on the cooked starch. When the food is acidic, *pepsin* and rennin act on the proteins and caseinogen. The food is quickly acidified in the pyloric end of the stomach, where peristaltic action is very marked, so that it acts like a mill, churning the food up and mixing it with the gastric juice. The food remains in the stomach for 30 minutes to 3 hours or more, according to the nature of the food and the muscularity of the individual stomach. A meal rich in carbohydrate but containing little protein, such as tea, toast and cake, will leave the stomach in half an hour. A good mixed meal, like a typical dinner, will remain for 2½ to 3 hours or more, though it may leave earlier or stay longer according to the tone and activity of the muscular coat.

The hydrochloric acid in the gastric juice serves several purposes:

- It gives the acid reaction required by the gastric enzymes
- It kills bacteria
- It controls the pylorus
- It stops the action of ptyalin
- It converts pepsinogen to pepsin.

The pylorus is normally contracted. When there is food in the stomach the gastric juice makes the contents gradually more acidic at the pyloric end. When it reaches a certain degree of acidity the pylorus relaxes and a little food passes into the duodenum (Fig. 19.8). The acidic food here causes the pylorus to close and the tone of the stomach wall drives food from the cardiac reservoir down to mix with the food in the pyloric end, making it less acidic. Gradually the food in the duodenum is made alkaline and that in the pyloric end of the stomach again becomes more acidic; this causes the pylorus to open again.

The churning action of the stomach serves to emulsify coarsely any fat which may be present and which the body heat will have melted. This converts the food into a greyish-white fluid called chyme. The gastric juice is very acidic and the lining of the stomach is normally protected by a layer of mucus, which it secretes. Any breakdown in this protection leads to erosion of parts of the lining of the stomach and also, sometimes, parts of the duodenum, causing

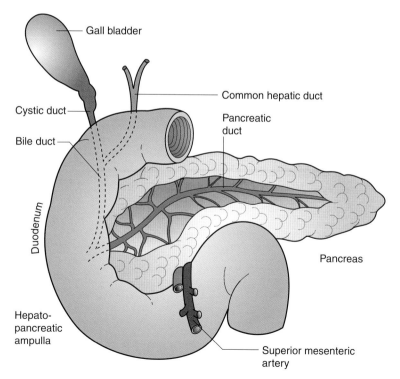

Fig. 19.8   The duodenum, pancreas and gall bladder.

peptic ulcers. Such ulcers are painful and if they are sufficiently eroded so as to expose blood vessels they can lead to loss of blood.

THE SMALL INTESTINE

The small intestine is a convoluted tube extending from the pyloric sphincter to its junction with the large intestine at the ileocaecal valve. It is about 6 m in length and lies in the central and lower parts of the abdominal cavity, usually within the curves of the large intestine (see Fig. 19.1). The small intestine consists of the *duodenum*, the *jejunum* and the *ileum*.

The duodenum is a short curved portion about 25 cm long and is the widest and most fixed part of the small intestine. It is roughly C shaped and curves round the head of the pancreas. Ducts from the *gall bladder, liver* and *pancreas* enter the medial aspect of the duodenum through the *hepatopancreatic ampulla*, which is guarded by a sphincter-like muscle. The duodenum plays a central role in controlling digestion, which will be considered below. When food from the stomach enters the duodenum, hormones are released, which simultaneously stimulate the release of bile from the gall bladder and pancreatic juice from the pancreas. These hormones – *cholecystokinin, secretin* and another called *gastric inhibitory peptide* – also inhibit movement in the stomach and the secretion of gastric juice. Also, the entry of food into the duodenum stimulates the nervous enterogastric reflex, which has an inhibitory effect on the stomach.

The jejunum is the name given to the upper two-fifths of the remainder of the small intestine; the lower three-fifths is called the ileum. Both are attached to the posterior abdominal wall by a fold of peritoneum called the *mesentery* (see Fig. 19.15).

The wall of the small intestine has the same four coats as the remainder of the alimentary tract:

- Serous coat, formed of peritoneum
- Muscular coat, with a thin external layer of longitudinal fibres and a thick internal layer of circular fibres
- Submucous coat, containing blood vessels, lymph vessels and nerves
- Mucous membrane lining.

The mucous membrane lining has three special features:

1. It is thrown into circular folds, which, unlike the rugae of the stomach, are permanent and are not obliterated when the

intestine is distended. They increase the area available for absorption (Fig. 19.9).

2. It has a velvety appearance due to the presence of fine hair-like projections called villi, each containing a lymph vessel called a lacteal, and blood vessels (Fig. 19.10).

3. It is supplied with glands of the simple, tubular type, which secrete intestinal juice.

The small intestine contains considerable amounts of lymphoid tissue. Solitary *lymphatic follicles* (Fig. 19.11) are found throughout the mucous membrane but are most numerous in the lower part of the ileum. Aggregated lymphatic follicles form circular or oval patches large enough to be seen with the naked eye. These nodules deal with

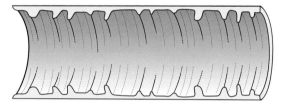

Fig. 19.9 Section of the small intestine, showing the puckered lining (circular folds).

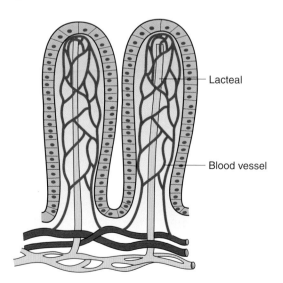

Fig. 19.10 Villi containing blood vessels and lacteal.

bacteria, which may be absorbed from the intestine; they become inflamed as a complication of such diseases as typhoid fever.

Functions of the small intestine

The functions of the small intestine are the digestion and absorption of food.

Digestion

Digestion is carried out by the pancreatic juice, bile (which emulsifies fats) and intestinal juice (Table 19.2). There are three phases of digestion, which work together:

- The cephalic phase
- The gastric phase
- The intestinal phase.

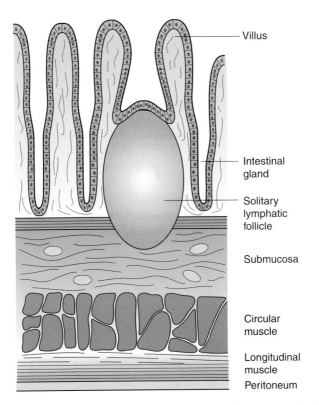

Villus

Intestinal gland

Solitary lymphatic follicle

Submucosa

Circular muscle

Longitudinal muscle

Peritoneum

Fig. 19.11   Section of the wall of the small intestine (highly magnified).

Table 19.2   Digestive processes

Source	Secretion	Changes
Saliva	Amylase	Cooked starch into maltose and dextrin
Gastric juice	Pepsin	Proteins into peptones
	Rennin	Coagulates milk (in babies)
	Hydrochloric acid	Stops action of amylase
		Converts pepsinogen to pepsin
Pancreatic juice	Trypsin	Proteins to polypeptides
	Amylase	All starch to maltose and dextrin
	Lipase	Fats to fatty acids and glycerol
Liver	Bile	Emulsifies fats
Intestinal juice	Enterokinase	Pancreatic trypsinogen to active trypsin
	Peptidase	Polypeptides to amino acids
	Maltase	Maltose ⎫
	Sucrase	Sucrose ⎬ to glucose
	Lactase	Lactose ⎭
	Lipase	Fats to fatty acids and glycerol

The *cephalic phase* occurs when we think of, see or smell food and it stimulates the release of gastric juice and begins movement in the stomach. This is the reason why our stomach apparently 'rumbles' when we are hungry. The *gastric phase*, as the name suggests, occurs when food is in the stomach, and the presence of food here stimulates the release of gastric juice and movement. The *intestinal phase* occurs when food enters the duodenum and the secretion and movement in the stomach is inhibited by the mechanisms described above, which involve both hormonal and neural mechanisms.

The *pancreatic* and *intestinal juices* are secreted by reflex through sensations concerning food, and also through a hormone called secretin produced by the lining of the intestine, as gastrin is secreted by the stomach wall. The juices are alkaline and make the food alkaline in reaction. This finely emulsifies the fat.

The pancreatic juice consists of water, alkaline salts and three enzymes acting on three different foodstuffs:

- Trypsinogen – turns peptones and proteins into amino acids when converted into active *trypsin* by *enterokinase*. If active trypsin were secreted by the pancreas, it would be able to digest the protein of the cells that form the gland and its ducts. Trypsin becomes active only when it mixes with the food and the intestinal juice within the bowel

- Amylase – turns starch (cooked and uncooked) into malt sugar (i.e. maltose)
- Lipase – splits fat into fatty acids and glycerol, after the bile has emulsified the fat, to increase the surface area.

The *bile* contains no enzymes but is rich in alkaline salts, which serve to *emulsify* and saponify fats, i.e. make soaps from them.

The intestinal juice contains water, salts and enzymes:

- Enterokinase – converts trypsinogen secreted by the pancreas to active trypsin
- Peptidase – acts on peptones and turns them into amino acids
- Maltase – turns maltose into simple sugar, such as glucose
- Sucrase – turns cane sugar (sucrose) into simple sugar
- Lactase – turns lactose into simple sugar
- Lipase – completes the conversion of fats to fatty acids and glycerol.

These juices are mixed with the food by peristalsis, the muscular action of the wall of the small intestine. The contractions occur first at one place and then at another and are followed by relaxation, having a kneading or churning effect and bringing the mucous lining into close contact with the contents of the gut.

Absorption

The absorption of proteins, carbohydrates and fats takes place almost entirely through the *villi* in the small intestine. Very little food is absorbed from the stomach as it is not yet sufficiently digested or, if absorbable, e.g. glucose and water, does not stay in the stomach but merely passes through it. Proteins in the form of amino acids, and carbohydrates in the form of simple sugar, are absorbed by the cells covering the villi and pass into the blood capillaries, being carried by the portal vein to the liver. Fats in the form of fatty acids and glycerol are absorbed by the cells covering the villi and built up again by them into droplets of fat. These pass into the lymph within the villi and are drained away by the lymphatic capillaries or *lacteals,* which are so named because the lymph they contain is made milk-like by the droplets of fat in suspension. The fat passes via the lymphatic vessels to the cisterna chyli and is carried up the thoracic duct into the blood stream.

THE LARGE INTESTINE

The large intestine extends from the end of the ileum to the anus (Fig. 19.12) and is about 15 m long. It forms an arch, which encloses most of the small intestine, and is divided into seven sections:

- Caecum
- Ascending colon
- Transverse colon
- Descending colon
- Sigmoid colon
- Rectum
- Anal canal.

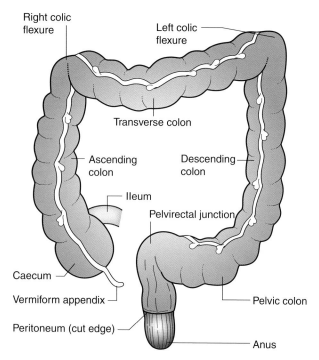

Fig. 19.12 The large intestine.

The *caecum* lies in the right iliac fossa. It is a dilated area with a blind lower end but is continuous above with the ascending colon and, where one passes into the other, the ileum opens into the caecum through the *ileocaecal valve* (Fig. 19.13). This valve is a sphincter and prevents the caecal contents passing back into the ileum. Taking food into the stomach initiates contraction of the duodenum and the rest of the small intestine, followed by the passage of the contents of the ileum into the caecum through the ileocaecal valve. This is called the *gastro-ileal reflex*.

The *vermiform appendix* is a narrow blind-ended tube that opens out of the caecum about 2 cm below the ileocaecal valve. It is usually about 9 cm long, though it can vary from 2 to 20 cm in length and can occupy a variety of positions within the abdomen. The submucous coat of the appendix contains considerable amounts of lymphoid tissue.

The *ascending colon* is about 15 cm in length and is narrower than the caecum. It ascends the right side of the abdomen to the

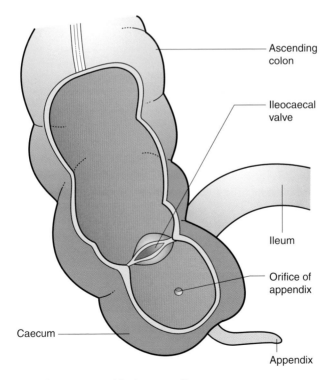

Ascending colon

Ileocaecal valve

Ileum

Orifice of appendix

Caecum

Appendix

Fig. 19.13   The caecum with the appendix.

undersurface of the liver where it bends forwards and to the left at the right colic flexure.

The *transverse colon* is about 50 cm in length and passes across the abdomen to the undersurface of the spleen in an inverted arch. Here it curves sharply downwards at the *left colic flexure.*

The *descending colon* is about 25 cm in length and passes down the left side of the abdomen to the inlet of the lesser pelvis, where it becomes the sigmoid colon.

The *sigmoid colon* forms a loop which is about 40 cm in length and lies within the lesser pelvis.

The *rectum* is continuous above with the sigmoid colon. It is about 12 cm long and passes through the pelvic diaphragm to become the anal canal.

The *anal canal* passes downwards and backwards to end at the *anus.* At the junction of the anus and the rectum the unstriped circular muscle becomes thickened to form the *internal anal sphincter,* which surrounds the upper three-quarters of the anal canal. The external anal sphincter surrounds the whole length of the anal canal; it is the tone of these sphincters that keeps the anal canal and the anus closed. The *external sphincter* can be contracted voluntarily to close the anus more firmly.

The wall of the large intestine has the same four layers as the remainder of the alimentary tract:

- The outer serous layer of peritoneum
- The muscular layer of external longitudinal and internal circular fibres. The longitudinal fibres form a continuous layer but in some places the layer is thickened to form three bands called taeniae coli. The bands are shorter than the other coats of the large intestine and produce the typical puckered or sacculated appearance. The sacculations are called the haustrations
- The submucous layer
- The lining of mucous membrane.

Functions of the large intestine

The functions of the large intestine are:

- To absorb water and salts
- To excrete faeces.

The material that enters the large intestine (Fig. 19.14) consists of water, salts, very little food material (as this has been digested and

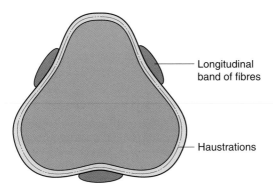

Longitudinal band of fibres

Haustrations

Fig. 19.14   Cross-section of the large intestine.

absorbed in the small intestine), cellulose (which is indigestible in humans) and bacteria. The bacteria are numerous; although they are mainly killed in the stomach, the alkaline reaction, food, warmth and moisture in the small intestine encourage their growth here. The material is in a very fluid state. In the colon, water and salts are quickly absorbed so that the fluid is rapidly turned into a paste containing the cellulose and bacteria, many of which die from lack of water and food. This paste forms the *faeces*, which consists of a fluid and a solid part, the latter being roughly 50% cellulose and 50% dead bacteria. The action of the bacteria in the large intestine is to ferment residual products in the faeces; one of these products is a range of substances called short-chain fatty acids (SCFAs). SCFAs are implicated in growth of the mucosa of the large intestine, and lower levels are produced on low fibre diets. At intervals, mass movements propel the faeces into the rectum, from which it is excreted.

Movements of the colon are similar to those in the small intestine but peristalsis occurs less frequently. A very strong wave of peristalsis occurs three to four times a day and moves material along into the pelvic colon.

Defecation, or the passage of faeces, occurs because of the movement of food residue from the pelvic colon into the rectum, which becomes distended. This distension causes a reflex contraction of the rectal muscles, which tends to expel the contents through the anus, though this depends on the relaxation of the external anal sphincter. Defecation is therefore a reflex action that can be inhibited voluntarily. The gastro-ileal reflex also causes emptying of the rectum. If defecation is delayed the sensation of fullness in the rectum passes off, more water is absorbed from the faeces through the rectal wall and constipation may develop.

THE PERITONEUM

The *peritoneum* (Fig. 19.15) is a serous membrane and in the male is a closed sac lining the abdomen. In the female the free ends of the uterine tubes open into the peritoneal cavity. The part that lines the abdominal wall is called the *parietal portion* of the peritoneum; the part that is reflected over the organs is the *visceral portion*. The

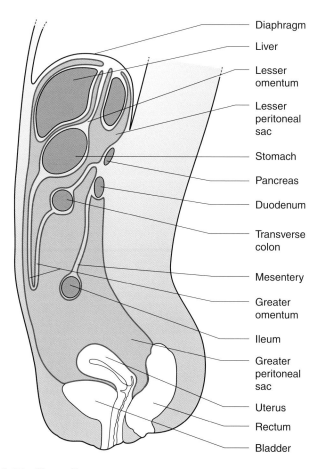

Fig. 19.15   The peritoneum.

two layers are in contact with each other but the potential space between them is named the *peritoneal cavity*. This cavity consists of the *greater peritoneal* sac and the *lesser peritoneal sac*.

Other specially named areas are the *greater omentum*, a double fold of peritoneum that descends from the lower border of the stomach and loops up again to the transverse colon. This helps to prevent the spread of infection from the organs into the peritoneum. The *lesser omentum* is a fold extending to the liver from the lesser curvature of the stomach and duodenum. The mesentery is a broad, fan-shaped fold of peritoneum connecting the coils of the small intestine to the posterior abdominal wall.

The functions of the peritoneum are:

- To prevent friction as the abdominal organs move against one another and against the abdominal wall, as its free surfaces are always moist with serum and are smooth and glistening

- To attach the abdominal organs to the abdominal wall, except in the case of the kidneys, duodenum and pancreas, which lie behind it. The ascending and descending colon are covered only on their anterior surface by peritoneum. This means that only the transverse or sigmoid colon can be brought on to the anterior abdominal wall for a colostomy operation

- To carry blood vessels, lymphatics and nerves to the organs; these run between the two folds of the membrane

- To fight infection, as the peritoneum contains many lymphatic nodes.

If the contents of the digestive tract come into contact with the peritoneum, a very dangerous condition called peritonitis can occur. In this condition, which is very painful, there is widespread inflammation of the peritoneum.

REGIONS OF THE ABDOMEN

For the purposes of description, the abdomen is divided into nine regions by two transverse and two upright lines (Fig. 19.16). In describing the positions of the organs, the regions in which they lie may be used, e.g. the stomach lies in the left hypochondriac, epigastric and umbilical regions. The kidneys are in the right and left lumbar regions, respectively. The caecum is in the right iliac fossa. The bladder rises when full into the hypogastric region.

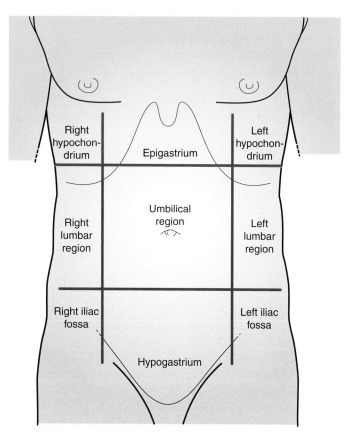

Fig. 19.16   Regions of the abdomen.

SELF-TEST QUESTIONS

- What are the functions of saliva?
- What are the layers of the digestive tract?
- Describe the phases of digestion.
- Which substances are digested in the small intestine?
- What are the sections of the large intestine?

Chapter 20

The liver, biliary system and pancreas

LEARNING OBJECTIVES

After reading this chapter you should understand:

- the structure and function of the liver
- the structure and function of the biliary system
- the structure and function of the pancreas

Due to their anatomical proximity and their related functions in digestion, the liver, biliary system and pancreas are usually considered together. This chapter looks at the structure of the liver and its function, particularly in the production of bile. The way in which the biliary system concentrates and subsequently releases bile is described and, finally, the structure and secretions of the pancreas are considered. The function of the pancreatic enzymes is the subject of Chapter 22.

THE LIVER

The liver is the largest gland in the body. It is situated in the upper right part of the abdominal cavity, occupying almost all of the right

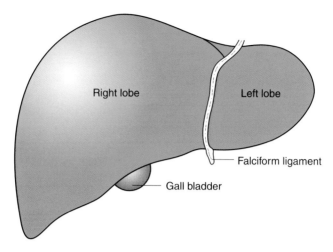

Fig. 20.1   General appearance of the liver.

hypochondrium and fitting under the diaphragm. It has two main lobes, the right lobe being much larger than the left (Fig. 20.1). The *right lobe* lies over the right colic flexure and the right kidney and the *left lobe* over the stomach.

Structure of the liver

The liver consists of a large number of hepatic *lobules* which appear hexagonal in shape (Fig. 20.2). Each is about 1 mm in diameter and has a small central *intralobular vein* (a tributary of the hepatic veins). Around the edges of the lobules are the *portal canals*, each containing a branch of the portal vein (interlobular vein), a branch of the hepatic artery and a small bile duct. These three structures together are known as the *portal triad*.

The lobules are composed of *liver cells*, which are large cells with one or two nuclei and fine granular cytoplasm. The liver cells are arranged in sheets, one cell thick, called hepatic laminae. These laminae are arranged irregularly to form walls with bridges of liver cells connecting adjacent laminae. Between the laminae are spaces containing small veins with many anastomoses between them and small bile ducts called canaliculi.

In the liver the portal vein brings blood rich in foodstuffs from the alimentary tract and the hepatic artery brings blood rich in oxygen from the arterial system. These divide into smaller vessels and

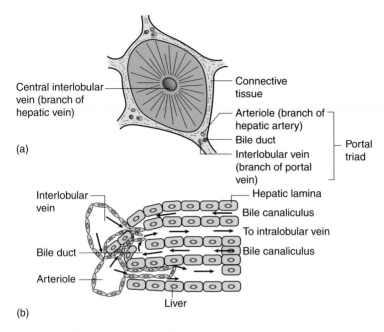

Fig. 20.2   (a) A liver lobule and (b) the arrangement of blood vessels and bile ducts.

provide a capillary network among the liver cells forming the hepatic laminae. This capillary network then drains into the small veins in the centre of each lobule, which supply the hepatic vein. These vessels carry away the blood from both the portal capillaries and the deoxygenated blood, which has been brought to the liver by the hepatic arteries as oxygenated blood.

Functions of the liver

The functions of the liver can be divided into three sections: metabolic; storage; and secretory:

- Metabolic functions:
  - stored fat is broken down to provide energy; this process is called desaturation
  - excess amino acids are broken down and converted to urea
  - drugs and poisons are detoxified
  - vitamin A is synthesized from carotene
  - the liver is the main heat-producing organ of the body

- the plasma proteins are synthesized
- worn-out tissue cells are broken down to form uric acid and urea
- excess carbohydrate is converted to fat for storage in the fat depots
- prothrombin and fibrinogen are synthesized from amino acids
- antibodies and antitoxins are manufactured
- heparin is manufactured
- Storage functions:
  - vitamins A and D
  - anti-anaemic factor
  - iron from the diet and from worn-out blood cells
  - glucose is stored as glycogen and converted back to glucose in the presence of glucagon as required
- Secretory functions:
  - bile is formed from constituents brought by the blood.

The formation of urea

The amino acids derived by the process of digestion from the protein food we eat are absorbed by the villi of the small intestine and brought by the portal vein to the liver. The amino acids required to make good the wear and tear of tissue and produce its growth are allowed to pass straight through the liver into the blood stream. Others are used to form the blood proteins. Any excess protein or second-class protein that is unsuitable for tissue building is broken up in the liver to form body fuel (composed of carbon, hydrogen and oxygen) and urea (a compound containing the nitrogen present in all proteins, which is incombustible, and therefore useless unless needed for the building of tissues.) *Urea* is a soluble substance, which the blood stream carries from the liver to the kidneys for excretion from the body.

The secretion of bile

The bile is a thick greenish-yellow fluid secreted by the liver cells, which is alkaline in reaction. The liver secretes on average about 1 litre of bile per day. The bile consists of water, bile salts and bile pigments. The *bile salts* give the bile its alkaline reaction and include both organic and inorganic salts. Among the former is cholesterol, which is the main ingredient of the typical gallstone. The *bile pigments* are derived from the haemoglobin of the worn-out red corpuscles, which are excreted from the body by this channel, giving the normal colour to the faeces. They are also present in the blood and colour the urine.

The functions of bile are that:

- It helps to emulsify and saponify fats in the small intestine by its alkalis; in this way the surface area is increased and the action of enzymes is increased

- It stimulates peristalsis in the intestine so that it acts as a natural aperient (i.e. laxative)

- It is a channel for the excretion of pigments and toxic substances from the blood stream, such as alcohol and other drugs

- It acts as a deodorant to the faeces, lessening their offensive odour. It is suggested that this may be due merely to the fact that a lack of bile means poor digestion of fat, so that fat remains in the intestine in excess, coating the other foods and preventing their digestion and absorption. As a result, undigested protein is attacked by bacteria and, decomposing, produces an excess of hydrogen sulphide, the gas that causes the smell of abnormal faeces, foul drains and rotten eggs.

THE BILIARY SYSTEM

The components of the biliary system are:

- The right and left hepatic ducts from the liver, which unite to form the common hepatic duct
- The gall bladder, which acts as a reservoir for bile
- The cystic duct, leading from the gall bladder
- The bile duct formed by the junction of the common hepatic and cystic ducts.

The gall bladder is a pear-shaped organ situated on the under-surface of the right lobe of the liver. From it the *cystic duct*, which is about 3–4 cm long, passes backwards and downwards to join the *common hepatic duct* and together they form the *bile duct* (Fig. 20.3). If the bile secreted by the liver is not required immediately for digestion it passes up the cystic duct into the gall bladder where it is both stored and concentrated. The capacity of the gall bladder is 30–60 mL but, because of its capacity to absorb water, the bile it contains

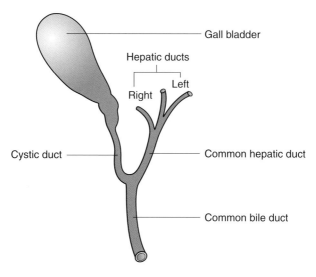

Fig. 20.3   The gall bladder and its ducts.

becomes increasingly concentrated. When fatty food enters the duodenum the sphincter at the entrance to the bile duct relaxes and bile stored in the gall bladder is driven into the intestine by contraction of the walls of the gall bladder.

From the list of its functions, it will be seen that the liver is essential to life. It is, however, able to undertake more work than is usually demanded of it and a considerable part may be destroyed by disease before death occurs from liver failure.

Jaundice

Jaundice describes any condition where bile enters the blood stream, thereby causing a characteristic yellowing of the skin and sclera of the eyes. Jaundice may be caused by liver disease, such as hepatitis or cirrhosis, where the liver cells break down and release bile into the blood.

If the outlet from the gall bladder via the common bile duct becomes blocked, bile will not flow into the small intestine. Instead, it flows back up the common hepatic duct and eventually enters the blood stream. This condition is known as *obstructive jaundice* and may be caused by gallstones, which can form in the gall bladder and travel down the cystic duct.

THE PANCREAS

The pancreas is a soft greyish-pink gland, 12–15 cm long, lying transversely across the posterior abdominal wall (see Fig. 19.8) behind the stomach. The head of the gland lies within the curve of the duodenum and the tail extends as far as the spleen. The body lies between these two. The *pancreatic duct* lies within the organ. It begins with the junction of the small ducts from the *pancreatic lobules* in the tail of the pancreas and runs from left to right through the gland, receiving ducts all the way. At the head of the pancreas the pancreatic duct is joined by the bile duct and they usually open together into the duodenum at the hepatopancreatic ampulla, though occasionally there are two separate ducts.

The pancreas is composed of lobules, each of which consists of one of the tiny vessels leading to the main duct and ending in a number of alveoli. The *alveoli* are lined with cells that secrete the enzymes trypsinogen, amylase and lipase, which have the following functions:

- Trypsinogen is converted into active trypsin by enterokinase, an enzyme secreted by the small intestine. In its active form trypsin converts peptones and proteins into amino acids
- Amylase turns starches (cooked and uncooked) into maltose (malt sugar)
- Lipase splits fats into fatty acids and glycerol after the bile has emulsified the fat, which increases the surface area.

Inflammation of the pancreas is called pancreatitis and can occur if the protein digesting enzymes come into contact with the tissue of the pancreas.

Between the alveoli, collections of cells are found forming a network in which there are many capillaries (Fig. 20.4). These cells are called interalveolar cell islets (*islets of Langerhans*) and they secrete a hormone that passes directly into the blood stream. The pancreas therefore has both a digestive and an endocrine function. Each islet consists of two types of cell, designated alpha and beta. Alpha cells form about 25% of the total number of islets and they produce a hormone (glucagon) which is secreted in response to a fall in blood sugar. *Glucagon* stimulates the conversion of glycogen to glucose, thus raising the blood sugar level. Beta cells form the remaining 75% of the islets and they secrete the hormone *insulin* in response to a rise in blood sugar level, e.g. after a meal. Insulin lowers the blood sugar level by stimulating the conversion of glucose to glycogen for

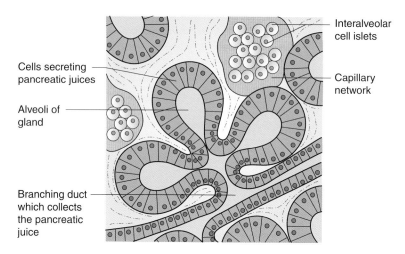

Interalveolar
cell islets

Cells secreting
pancreatic juices

Capillary
network

Alveoli of
gland

Branching duct
which collects
the pancreatic
juice

Fig. 20.4   The minute structure of the pancreas, showing the saccules, which secrete the pancreatic juice, and the islet cells, which secrete insulin.

storage and by increasing cellular uptake of glucose. It will be apparent therefore that the blood sugar level is maintained by a balance between the two hormones because they both affect carbohydrate metabolism. As the metabolism of proteins and fats is closely related to carbohydrate metabolism a disturbance in one will affect the others.

A deficiency of insulin results in *diabetes mellitus*. The blood sugar level rises above the renal threshold and glucose is lost in the urine. Because the cells cannot use glucose, there is an accumulation of ketone bodies as a result of the breakdown of fatty acids, which causes acidosis and may lead to coma and death if untreated.

SELF-TEST QUESTIONS

- What does the gall bladder do?
- What is the function of bile?
- Name the digestive enzymes of the pancreas.

Chapter 21

Nutrition

LEARNING OBJECTIVES

After reading this chapter you should understand:

- which are the essential foodstuffs
- the different types of sugar in the diet
- what happens when the diet is deficient in certain vitamins
- current nutritional guidelines

Good health is dependent on satisfactory nutrition, and this is in turn dependent on plentiful supplies of the foods necessary for a healthy life. Food is one of the essential needs of the body. Substances that can serve as food for the body are those that it can use as fuel for combustion, or building material for the repair and growth of tissues. Fuel is required to produce the energy for the activities of every living thing and to maintain the heat at which the individual exists. Building materials are necessary to repair the body tissues since they are constantly active and being worn out by their activities. In addition, in infants and children, extra building material is required to build up the new tissues needed for the processes of growth. Fuel supplies and building materials alone are, however, not enough. Certain other substances are necessary to enable the tissues to use the building materials and fuel supplies, and these are known as vitamins.

There are six essential foodstuffs with which the body must be constantly supplied through the foods we eat:

- Proteins
- Carbohydrates
- Fats
- Water
- Mineral salts
- Vitamins.

Every article of food contains one or more of these foodstuffs and is only of value as food because it does contain them. The proteins, water and salts are the body-building foods; the carbohydrates and fats are essentially the fuel foods, though the body can and does use protein also as fuel, if more is taken in than is required for body building, or if there is a lack of other fuel, as in starvation. The vitamins and certain salts act as regulators of tissue activity so that, although vitamins are of no use either as fuel or as body builders, the nutrition of the tissues suffers if they are not present in sufficient quantities in the food supply, and diseases appear, which can be prevented and cured by ensuring that the vitamins are present in satisfactory quantities in the diet.

To be of use to the body, these foodstuffs must be digested and absorbed. Food must therefore be of such a nature that it can be digested, i.e. broken down by digestive juices into substances that can pass into the blood stream and be carried to the various tissues for their use. Proteins, carbohydrates and fats are complicated compounds found in plant and animal matter, and require digestion.

Water and mineral salts are simple inorganic substances, therefore they can be absorbed without digestion and they enter into the composition of all animal and plant matter. In fact, all living matter consists largely of water. The inorganic salts, absorbed from the soil or water, are built up by living things into organic salts, which are an essential part of all plants and animals. (See Chapter 2 for more details of the chemical structure of nutrients.)

PROTEINS

Proteins are the most complicated of the foodstuffs. They consist of carbon, hydrogen, oxygen, nitrogen and sulphur, and usually phosphorus. They are often spoken of as the nitrogenous foodstuffs as they are the only ones that contain the element nitrogen. They are essential for the building up of living protoplasm as this also consists of the elements carbon, hydrogen, oxygen, nitrogen and sulphur. They are found in both animal and plant matter although the animal proteins are the most valuable to the human body as building material since they are similar to human protein in composition. On the other hand, plant proteins are cheaper: they are more useful as body fuel than as body builders but do provide, at a lower cost, some of the amino acids that the body needs for tissue building. The sources of protein are:

- Animal:
    - eggs, containing albumin
    - lean meat, containing myosin
    - milk, containing caseinogen and lactalbumin
    - cheese, containing casein
- Plant:
    - wheat and rye, containing gluten
    - pulse foods (e.g. peas and beans) containing legumin.

All proteins are made up of simpler substances known as amino acids. There are about 20 of these amino acids but each protein contains only some of these. The amino acids are like letters from which many words can be made, each word being a different combination of letters. The protein of each different type of animal or plant is a different combination of amino acids. Ten essential amino acids are found in human protein; these are amino acids that the body cannot build up for itself. Proteins that contain all ten are called complete proteins, e.g. albumin, myosin and casein.

Proteins that do not contain all the ten *essential amino acids* are called incomplete proteins, e.g. gelatin, which is contained in all fibrous tissue, and is extracted from bone and calves' feet in the making of soups and jelly. Animal proteins, such as those of eggs, milk and meat, not only contain all the ten amino acids the body needs but contain all of them in good proportion; these are called first-class proteins and are the best building material for the body tissues. Plant proteins, such as gluten and legumin, contain only slight quantities of one or more of the ten amino acids essential to the body and are therefore called second-class proteins as they are not such good building material. Some first-class animal protein should always be included in the diet.

CARBOHYDRATES

Carbohydrates include sugar and starch. They consist of carbon, hydrogen and oxygen, and always contain hydrogen and oxygen in the same proportions as in water, i.e. twice as much hydrogen as oxygen. They are the chief sources of body fuel, being most easy to digest and absorb and most readily burnt in the tissues, being broken down into carbon dioxide and water. They are obtained chiefly from plant foods:

- Sources of starch:
  - cereals, e.g. wheat, rice and barley
  - tubers and roots, e.g. potatoes and parsnips
  - pulse foods, e.g. peas, beans and lentils
- Sources of sugar:
  - sugar cane (sucrose)
  - beetroot and all sweet vegetables and fruits, e.g. grape sugar or glucose
  - honey
  - milk, containing lactose or milk sugar.

Sugars are of three types:

- Simple sugars (monosaccharides), such as glucose or grape sugar (formula $C_6H_{12}O_6$)
- Complex sugars (disaccharides), such as cane sugar or sucrose, and milk sugar or lactose (formula $C_{12}H_{22}O_{11}$)
- Polysaccharides, the most complex carbohydrates, e.g. starches such as in potatoes, cereals and root vegetables.

Starch differs from sugar in that it is insoluble in water. Plants store sugar in the form of starch to prevent it from escaping in solution into the water in the soil in which they live. (The formula for starch is $(C_6H_{10}O_5)_n$, a polysaccharide; n stands for different numbers in the different starches of various plants.) All carbohydrates are reduced to monosaccharides before they can be absorbed from the digestive tract.

FATS

Fats, like carbohydrates, consist of carbon, hydrogen and oxygen, but do not contain as much oxygen in proportion to the hydrogen present. They also serve as body fuel. They are the best source of fuel from the point of view that 1 g of fat produces twice as much energy as 1 g of sugar. On the other hand, they are not so easy to digest and absorb and not so readily burnt in the tissues. Fats can only be completely broken down to form the final products of metabolism, carbon dioxide and water, if they are metabolized with sugar. If sufficient sugar is not metabolized with them, the breakdown of fat is incomplete and acid or acetone bodies are formed in the tissues. These acetone bodies cause fatigue in the muscles and, if present in large quantities, alter the pH of the blood, causing a condition known as acidosis, which may lead to coma and death. Severe acidosis is only likely to occur in diabetic patients, who are unable to metabolize sugar, and in starvation, when the small quantity of sugar that the body can store has been used up and the large quantities of fat that the body can store are forming the main source of body fuel.

Fats are obtained from both animal and plant matter. The main sources are:

- Animal:
    - fat meat and fish oils
    - butter
    - milk and cream
- Plant:
    - nut oil, contained in margarine
    - olive oil.

Fats are compounds of glycerol and fatty acids. Different fats contain different acids, e.g. fat meat contains stearic acid, butter contains butyric acid, olive oil contains oleic acid, etc.

Animal fats are more expensive than plant fats but are more valuable as food because they contain the fat-soluble vitamins A and D provided the animals have been exposed to sunlight and not kept in stalls. However, plant fat can be made equally valuable by the addition of vitamin D, and all margarine, whether manufactured from fish oil or plant oil, is so treated today to ensure the supply of these vitamins. In this way the population has been protected from deficiency diseases due to lack of these fat-soluble vitamins.

WATER

Water is a simple compound of hydrogen (two parts) and oxygen (one part). It forms two-thirds of the body and is present in most of the foods we eat. Lean meat is three-quarters water, milk contains 87% water, while cabbage contains as much as 92% water. In addition to the water contained in food the body needs 2–3 litres of water every day. Water is required for many purposes:

- The building of body tissues and body fluids
- The excretion of waste products
- The making of digestive and lubricating fluids
- The cooling of the body by the evaporation of sweat

Fluid and electrolyte balance is discussed in Chapter 23.

ELECTROLYTES AND MINERALS

Salts and the ions produced from them are constantly taken into and lost from the blood. Salts are contained in all foodstuffs but added salt, used in cooking and eating, is necessary to maintain the normal electrolyte balance and replace the salt lost in urine and sweat; 3–4 g of sodium chloride, or common salt, is needed daily.

The quantities of the various salts are estimated as millimoles per litre (mmol/L) of the ions they produce. In plasma there are normally 155 mmol/L of negatively charged ions, balanced by 155 mmol/L of positively charged ions. The bulk of these are provided by chloride (102 mmol/L) and sodium (143 mmol/L). These figures are given for reference, as the nurse will see them in laboratory reports to check for deficiencies or excess of sodium, potassium chloride, bicarbonate, etc. in the blood. These may result from deficiency in intake or excessive or reduced loss, especially in abnormal conditions affecting the kidney, or from heavy sweating or fever. Table 21.1 shows some electrolytes that are required in very small amounts and where they can be obtained, with reference values.

MINERAL SALTS

Salts are produced by the action of an acid on a mineral, which is termed the base of the salt. For example, sodium chloride (common salt) is produced by the action of hydrochloric acid on sodium; calcium lactate is produced by the action of lactic acid on calcium. Various salts and the ions split off from them are required by the body; every tissue and fluid in the body contains them. Chlorides, carbonates and phosphates of sodium, potassium and calcium are particularly important. Salts are required for body building and are also regulators of tissue activity. They provide the electrolytes, capable of carrying positive and negative electric charges in the body fluids (see Chapter 1). A correct balance of electrolytes is essential for normal activity in the body tissues and fluids.

Table 21.1   Elements required in small amounts

Element	Symbol	Rich sources[a]	Reference nutrient intake for adults (mg/day)	(μg/day)
Iron	Fe	Treacle, liver, black pudding	8.7–14.8	–
Zinc	Zn	Oysters,[b] meat, whole grains legumes	7–9.5	–
Manganese	Mn	Plant foods (tea)	1.4	–
Copper	Cu	Green vegetables, liver	1.2	–
Molybdenum	Mo	Plant foods	[d]	–
Fluoride	F	Drinking water, tea	[e]	–
Iodide	I	Seafood[c]	–	14.0
Selenium	Se	Fish	–	60–70
Chromium	Cr	Widespread	–	[f]

[a] Some foods that are less rich may contribute a large proportion of the dietary intake if they are eaten in large quantities. Cereals, for example, can be a significant dietary source of iron because of the amounts consumed each day.
[b] Oysters are exceptionally rich in zinc (10 to 20 times as much as in meat, legumes, etc.).
[c] In some countries (USA, Switzerland, New Zealand but not the UK) iodine is also obtained from table salt which, in these countries, is iodized.
[d] No RNI set but safe intakes in adults between 50 and 400 μg per day.
[e] No RNI set but an upper limit on intakes of infants and young children of 0.05 mg/kg/day is suggested.
[f] No RNI set. A safe adequate level for children of 0.1–1.0 mg/kg/day and 25 μg/day for adults.

From Kindlen S (ed) 2003 Physiology for health care and nursing, 2nd edn. Churchill Livingstone, Edinburgh.

Sodium is present in all tissues – the body contains it in the form of sodium chloride to the extent of 9 g to 1 litre (i.e. 0.9%). It is present in the same concentration in all tissue fluid. Sodium carbonate and sodium phosphate are also always present in the blood and tissues. The carbonates give the alkaline reaction to the blood and form the alkali reserve, neutralizing the carbonic acid produced by fuel combustion. The phosphate is the carrier of the acids produced by the breakdown of the body-building foods and carries them to the kidneys, by which they are excreted. Sodium is obtained from food, particularly animal foodstuffs, and from *salt* used in cooking.

Potassium is present in all tissue cells, where it replaces the sodium of blood and tissue fluid as the base and the source of the positively charged ions. It is obtained from food, particularly plant foodstuffs.

Calcium is present in all tissues, particularly in the bones, teeth and blood, and is necessary for the normal functioning of nerves. It is obtained chiefly from milk, cheese, eggs and green vegetables, and to some extent from hard water, though this inorganic form is less valuable than the organic salts in plant and animal foods. Adults require 400–500 mg daily.

Iron is essential for the formation of the haemoglobin of red blood cells. It is obtained from green vegetables (particularly spinach and cabbage), egg yolk and red meats. Men require 10 mg daily, women 10–15 mg.

Phosphorus is also needed for the building of the body tissues. It is obtained from egg yolk, milk and green vegetables.

Iodine is required for the formation of the secretion of the thyroid gland. It is obtained from seafood and is present in green vegetables, which have derived it from the fine spray of sea water in the air, which is carried inland by winds. Calcium, iron and iodine are the only minerals likely to be insufficient. The others are present in adequate amounts in the diet.

VITAMINS

Vitamins are substances that are also essential to normal health although they are of no value to the body as fuel or building material. Without them, diseases occur, which are known as deficiency diseases. Vitamins are present in small quantities in living foodstuffs and are required only in minute traces each day. Their discovery dates back to the period following the First World War (1914–1918) and, as their composition was at first unknown, they were named after the letters of the alphabet. Today, they can largely be made synthetically.

A number of vitamins have now been identified but experimental work still continues. The chief vitamins are vitamin A, the vitamin B complex, vitamin C, vitamin D, vitamin E and vitamin K. Tables 21.2 and 21.3 show the fat-soluble and water-soluble vitamins, respectively, giving their reference values and sources.

Vitamin A

Vitamin A is present in animal fats, being fat soluble. In carrots and green vegetables, and in all yellow fruits, a substance called carotene

Table 21.2 Fat-soluble vitamins

Vitamin	Alternative or equivalent name	Major sources	Reference nutrient intake[a]
A	Retinol (β-carotene)[b]	Butter, liver, fish liver oil, carrots	350–700 µg
D	Calciferol	Action of sunlight on skin, fish liver oil, egg yolk	8.5–10 µg
E	Alpha tocopherol	Vegetable oils, eggs	[c]
K	Menaquinone	Green leafy vegetables, bacterial metabolism (gut)	1 µg/kg/day (adults)[d] 10 µg/day (infants)[d]

[a] Dietary Reference Values for Food Energy and Nutrients for the United Kingdom. Report of the Panel on Dietary Reference Values of the Committee on Medical Aspects of Food Policy (1991) HMSO, London.

[b] Some foods such as carrots contain β-carotene which is converted to vitamin A in the body. For this reason the vitamin A content of foods is quoted as 'retinol equivalents' in order to include this source of vitamin A.

[c] The requirement for vitamin E is partly determined by the polyunsaturated fatty acid content of the diet. Signs of clinical deficiency are unlikely to arise.

[d] Production of vitamin K by intestinal bacteria makes estimation of dietary requirement difficult.

From Kindlen S (ed) 2003 Physiology for health care and nursing, 2nd edn. Churchill Livingstone, Edinburgh.

is found, which, like the green colouring matter chlorophyll, is formed under the influence of sunlight. This is a precursor of vitamin A, and animals, including humans, can turn the carotene present in their food into vitamin A within their bodies.

Lack of vitamin A causes stunted growth and lowered resistance to infection. The mucous membranes, particularly, are unhealthy and easy prey to bacteria when the diet is deficient in vitamin A. The conjunctiva of the eye are affected and a form of conjunctivitis occurs, known as *xerophthalmia*, in which the conjunctiva loses their transparency, and a horniness or cornification develops in it. The retina are also affected and night blindness (i.e. the inability to see at night) develops – a symptom that became apparent in the blackout during

Table 21.3   Water-soluble vitamins

Vitamin		Other name	Rich sources[a]	Reference nutrient intakes[b] (mg/day)	Reference nutrient intakes[b] (μg/day)
B group	B_1	Thiamin	Yeast	0.8–1	–
	B_2	Riboflavin	Yeast, liver	1.1–1.3	–
	B_3	Niacin	Yeast	1.2–1.8	–
	B_6	Pyridoxine	Liver	1.2–1.5	–
	B_{12}	Cobalamin	Liver, kidney, eggs	–	1.2–1.5
		Folic acid	Vegetables, yeast, liver	–	200
		Pantothenic acid	Liver	[c]	–
		Biotin	Egg yolk	–	[d]
C		Ascorbic acid	Fruit, vegetables	40	–

[a] Foods that are less rich may contribute a large proportion of the dietary intake because of the amounts consumed per day. Cereal-based foods (such as breakfast cereals) and milk, for example, are good sources of the B vitamins.
[b] Range of recommended values from several countries (Australia, UK, USA) for adults.
[c] No biochemical method accepted for determining pantothenate status in humans. No signs of deficiency observed in UK on intakes between 3 and 7 mg/day.
[d] No RNI set because of limited evidence; intakes of 10–20 μg/day considered safe and adequate.

From Kindlen S (ed) 2003 Physiology for health care and nursing, 2nd edn. Churchill Livingstone, Edinburgh

the Second World War when some people found they could not see as well as others in the dark because their diet had not been satisfactory.

Vitamin B

The vitamin B complex consists of a number of factors, although originally it was thought to be a single substance. These factors were confused because they had a similar distribution, being found particularly in the husks and germs of cereals and pulses, and in yeast and yeast extracts. They are also present to a lesser extent in vegetables, fruit, milk, eggs and meat. White flour and the bread, cakes and pastries made from it do not contain these factors, nor do polished rice and barley. Hence whole-grain bread and wholemeal flour have a food value that white bread and its flour lack and where these latter form a large part of the diet, a subnormal state of health may be present, or even deficiency disease, as occurred in many of the prisoner-of-war camps during the Second World War. The chief factors in the vitamin B complex are now considered.

Vitamin B_1 (aneurine or thiamin). This is essential for carbohydrate metabolism and controls the nutrition of nerve cells; a marked deficiency of it leads to *beri-beri*, in which there is inflammation of the nerves, causing paralysis and a loss of tone and activity in the intestinal muscles, with constipation, while the patient complains of loss of appetite and a burning sensation in the feet.

Vitamin B_2 (riboflavin). This is essential for the proper functioning of cell enzymes.

Vitamin B_3 (nicotinic acid). This is essential for carbohydrate metabolism. A lack of this vitamin causes pellagra, which leads to skin eruptions, gastrointestinal changes and mental changes.

Vitamin B_6 (pyridoxine). This is believed to be necessary for protein metabolism.

Vitamin B_{12} (cyanocobalamin). This is the anti-anaemic substance or factor absorbed by the villi of the small intestine and stored in the liver. It is satisfactorily absorbed only in the presence of an intrinsic factor produced by the lining of the stomach and hydrochloric acid. Vitamin B_{12} is essential for the proper development of red cells in the red bone marrow. Lack of it, or of its absorption, causes *pernicious anaemia*.

Folic acid. This is part of the vitamin B complex. It is also necessary for the maturation of red blood cells.

Vitamin C

Vitamin C (ascorbic acid) is water soluble and is found in fresh fruit, particularly citrus fruits (e.g. oranges, grapefruit and lemons) and in green vegetables and potatoes. It is important in tissue respiratory activity, wound repair and resistance to infection, and affects the condition of capillary walls, which become abnormally fragile if it is not present in plentiful supply in the diet. Lack of it causes *scurvy*, hence it is also called the antiscorbutic vitamin. It is particularly readily destroyed by heat therefore some fresh fruit or salad should be included in the daily diet. Cabbage and other greens provide a very rich source of vitamin C and are comparatively cheap: eaten raw, as salads, in a finely shredded form, they are very palatable and most valuable, provided the cabbage is fresh and crisp. Cooking for a long time lessens the content of vitamin C and, for this reason, cabbage should be finely shredded and cooked in boiling water only until tender, i.e. for about 10 to 15 minutes.

Under normal circumstances the eating of a good mixed diet, containing plenty of fresh fruit and vegetables, makes the taking of vitamin C in tablet form unnecessary and inadvisable.

Vitamin D

Vitamin D is fat soluble and is found with vitamin A in animal fats, provided the animal has been in the sun; cod liver oil and halibut liver oil are very rich in it. Halibut liver oil is more expensive as the supplies are less plentiful but it is richer in content, so it is useful for those who cannot digest cod liver oil. Vitamin D can also be made in the skin by the action of ultraviolet (UV) rays on the ergosterols present; further, it can be manufactured by subjecting the sterols in fats to the action of UV rays. The product of this process, calciferol, can be obtained in tablet form. Calciferol is identical with the vitamin D present in animal fat and can be used in illness and in emergencies as a substitute for it. Vitamin D is essential for the development of bone and teeth, affecting the use of calcium and phosphorus in the body. Lack of vitamin D causes *rickets*, hence it is also known as the antirachitic vitamin. This disease was common among the poor of industrial areas who ate margarine in place of butter, especially in places like the UK, where there is not a great deal of sunshine. Vitamin D is now included in the manufacturing of margarine and this has played a part in reducing rickets in the British population.

Vitamin E

Vitamin E is present in vegetable oils. It is present in cereals but little is known of its importance in human beings.

Vitamin K

Vitamin K is fat soluble and can be obtained from green vegetables and liver. People who have a deficiency of this vitamin in their diet show a tendency to haemorrhage since vitamin K helps to form prothrombin, which is essential for blood clotting. Vitamin K is synthesized in the intestines by bacterial action.

The quantities of the various vitamins required in the normal diet are estimated in milligrams; pure vitamins can be given in measured quantities on prescription.

NON-STARCH POLYSACCHARIDES

'Non-starch polysaccharides' is the nutritional term for the commonly used terms 'roughage' or 'dietary fibre'. They are present in large quantities in fruits, vegetables, beans and whole-grain cereals. Non-starch polysaccharides cannot be digested in the small intestine in humans and pass into the large bowel, where they are fermented by bacteria. They increase the bulk and transit time of the stool and therefore help to prevent constipation.

DIET

A daily energy intake for healthy adults is in the region of 1180–2550 kcal. This requirement may rise for people who are very active. It is the way in which this energy is obtained which is presently the focus of dietary advice. In 1994 the Committee on Medical Aspects of Food Policy (COMA) published guidelines with respect to the intake of dietary fat, carbohydrate and other nutrients by healthy groups in the UK population.

The COMA recommended that fat intake should be less than 35% of dietary energy intake and that the intake of saturated fats should be no more than 10% of energy intake. It was recommended that protein intake should comprise between 10 and 15% of energy intake (usually about 50 g per day) and carbohydrate intake should be 50–55% of dietary energy intake (primarily in the form of complex carbohydrates). Further, it was considered that non-milk sugars (e.g. glucose) should contribute no more than 10% of energy intake.

These guidelines can be achieved by substituting fish, poultry (without skin), lean meats and reduced fat dairy products for fatty meats and whole-milk dairy products. In addition, intake of oils, fats, eggs, fried foods and other high-fat foods should be limited. Carbohydrates should be consumed in complex forms, such as starch rather than in refined forms such as glucose. Specifically, it is recommended that five portions of vegetables and fruit (including some vegetables, some fruit and some salad) should be eaten each day.

Salt intake should not be too high (about 6 g per day) and this can be achieved in part by not adding salt to food at the table and limiting its use in cooking. Potassium intake can be obtained from consumption of a good intake of fruit and vegetables and should be about 3·5 g per day. An adequate intake of calcium can be achieved by consumption of dairy products, cereals and some vegetables and fish.

Vitamin and mineral supplements are not usually considered necessary if healthy eating guidelines are followed. However, supplements may be advised in some circumstances, for example, women who are planning a pregnancy should take folic acid supplements. It is emphasized that the best way to maintain a desirable body weight is to balance dietary intake with adequate physical activity.

SELF-TEST QUESTIONS

- What are the essential foodstuffs?
- Name the different types of sugar in the diet.
- What is the function of calcium in the body?
- Select a vitamin and explain what happens when it is deficient in the diet.
- What value are vegetables in the diet?

Chapter 22

Digestion and metabolism

CHAPTER CONTENTS

LEARNING OBJECTIVES

After reading this chapter you should understand:

- the digestion of carbohydrates, fats and proteins
- the main features of metabolism

The anatomy of the digestive tract was considered in Chapter 19. The aim of the present chapter is to consider the chemical breakdown (digestion) of individual foodstuffs and their subsequent use in metabolism to provide energy and structural components for the body.

The digestion of the major sources of nutrition – carbohydrate, fat and protein – is dealt with in the following sections, and the interrelated nature of their metabolism in a subsequent section (Figs 22.1 and 22.2).

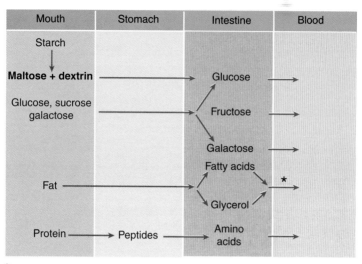

* Via lymphatic system

Fig. 22.1 Digestion and absorption of carbohydrate, fat and protein.

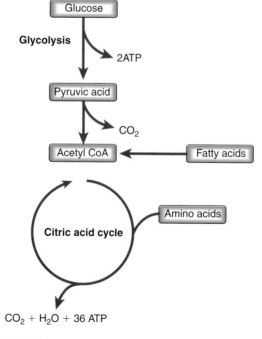

Fig. 22.2 Metabolism.

DIGESTION OF CARBOHYDRATES

Carbohydrate is present in the diet in several forms: as polysaccharides, disaccharides and monosaccharides. The polysaccharides that can be metabolized are starch (from plants) and glycogen (from animal sources). Both these polysaccharides are polymers (see Chapter 2) of glucose. The major polysaccharide is starch and the digestion of this begins in the mouth where *salivary amylase* begins to break it down into shorter polymers of glucose such as maltose and dextrin.

The action of *amylase* on starch is probably very short as food in the mouth is quickly swallowed and, when the food reaches the stomach, the hydrochloric acid inactivates the amylase. Some starch digestion may continue inside balls of food but, by the mixing action of the stomach movements, all the amylase is eventually inactivated.

There is no digestion of disaccharides and monosaccharides in the mouth or in the stomach. The main disaccharides present in the diet are sucrose, maltose and lactose (from milk). The composition of these disaccharides is:

- Sucrose: glucose–fructose
- Maltose: glucose–glucose
- Lactose: glucose–galactose.

The monosaccharides in the diet are glucose and fructose.

Once the contents of the stomach have passed into the duodenum the digestion of carbohydrates continues. In the duodenum, *pancreatic amylase* continues to break starch down into smaller polysaccharides and eventually into the disaccharide maltose. The intestinal juice contains three enzymes, maltase, lactase and sucrase, which complete the digestion of disaccharides into glucose, fructose and galactose. All the monosaccharides are subsequently absorbed in the small intestine from where they enter the blood stream and can then be metabolized in the body.

DIGESTION OF FATS (TRIGLYCERIDES)

Fats are not digested in the mouth or in the stomach, through which they pass as fat globules. The process of fat digestion begins in the duodenum where the physical properties of the fat globules are changed by the *emulsifying action of bile*. This breaks down the globules and increases the surface area of fat which can come into contact with the enzyme lipase from the pancreas. *Lipase* breaks fats down into fatty acids and glycerol, which are absorbed from the small intestine.

In the cells of the small intestine, fat is recreated and formed into small particles, with protein, called *chylomicrons*, and these are absorbed into the lymphatic system of the small intestine and taken to the blood stream. From the blood stream, they are taken to the tissues of the body, e.g. liver and muscle, where they are metabolized.

DIGESTION OF PROTEINS

Protein is not digested in the mouth. In the stomach, the action of hydrochloric acid changes the physical properties of protein, making it easier to digest. The process of protein digestion begins in the stomach under the action of the enzyme pepsin. *Pepsin* is produced in an inactive form called pepsinogen, which is activated to pepsin by the action of hydrochloric acid. Pepsin breaks protein macromolecules down into smaller fragments called peptides.

The process of protein digestion continues in the duodenum where the enzymes *trypsin* and *chymotrypsin* also act to break the remaining protein and peptides down into smaller fragments. Both these enzymes are produced by the pancreas in inactive forms, trypsinogen and chymotrypsinogen, which are activated in the alkaline juice of the pancreas. This alkaline juice stops the action of pepsin.

The small intestine contains further enzymes which complete the digestion of protein down to its constituent amino acids. These enzymes are *aminopeptidase* and *carboxypeptidase*, which digest single amino acids from the ends of the protein fragments. The enzyme dipeptidase breaks down dipeptides (composed of two amino acid molecules) into single amino acids. Amino acids are absorbed from the small intestine into the blood stream from which they are delivered to tissues for metabolism.

METABOLISM

Metabolism is composed of two processes – *anabolism* and *catabolism* – which work simultaneously to build up and break down molecules, respectively. The two 'phases' of metabolism are entirely dependent upon one another. The functions of metabolism are:

- The provision of energy in the form of adenosine triphosphate (ATP)
- The exchange of carbon between different types of molecule.

All the energy-using processes of the body, such as movement, are dependent upon metabolism, as are the synthetic processes whereby large molecules such as protein are made for use in muscle and other tissues.

The citric acid cycle

The citric acid cycle is often called the 'hub' of metabolism. The breakdown of nutrients to provide energy, and the exchange of carbon for synthetic purposes, is accomplished by this unique metabolic pathway.

The breakdown of glucose, and the use of glycerol in metabolism, is accomplished by the process of glycolysis. This produces pyruvic acid and two molecules of ATP. The pyruvic acid is converted to a substance called acetyl coenzyme A (i.e. acetyl CoA) and this is further broken down in the citric acid cycle to produce carbon dioxide, water and 36 molecules of ATP.

Fatty acids, which result from the breakdown of fats, are broken down by a process that produces acetyl CoA. Acetyl CoA is further broken down by glycolysis to produce carbon dioxide, water and ATP. The amount of ATP produced depends on the particular fatty acid that is being broken down, and it should be noted that fat produces more than twice the amount of ATP per gram as does carbohydrate and is, therefore, a much better source of energy.

Amino acids can be used to produce energy by being broken down in the citric acid cycle. Different amino acids enter this pathway at different points.

Due to the central role of the citric acid cycle in the metabolism of carbohydrate, fat and amino acids, it is possible to exchange carbon between molecules. For instance, some amino acids can be made from molecules in the citric acid cycle for subsequent use in protein synthesis, and glucose can be made from some amino acids in a process called gluconeogenesis. Note, however, that fatty acids cannot be used to make glucose as the conversion of pyruvic acid to acetyl CoA cannot be reversed.

SELF-TEST QUESTIONS

- What is starch broken down into in the mouth?
- How do fats get from the intestine into the blood?
- Describe protein digestion and where it takes place.
- What does the citric acid cycle achieve that glycolysis does not achieve?

Chapter **23**

Fluid balance and the urinary system

LEARNING OBJECTIVES

After reading this chapter you should understand:

- fluid balance
- the structure and function of the urinary system
- the functions of the kidneys
- the production of urine
- micturition

Maintaining adequate hydration and the correct distribution of fluids between the compartments of the body is vital. This chapter will look at what the normal distribution of fluid between the intracellular and extracellular compartments of the body should be. The urinary system plays a key role in ensuring that excess fluid is removed from the body. It also has an excretory function, which requires it to be constantly removing soluble toxic material from the body in the form of urine. This can be achieved even when fluid has to be conserved. This chapter looks at the anatomical structure of the urinary system and considers its ability to maintain fluid balance and respond to different fluid requirements of the body.

FLUID AND ELECTROLYTE BALANCE

Water is one of the essential foodstuffs but is a simple substance which can be absorbed and used in the body without chemical change. It enters into many of the metabolic changes that occur in the body, combining with proteins, carbohydrates and fats in digestion and being split off from them when they are used as fuel to produce energy. It is general today to consider the balance rather than the metabolism of water and salts, and the electrolytes that the salts form in the body.

Water forms the greater part of the body cells and the body fluids. Roughly two-thirds (more accurately 60%) of the body weight consists of water. This proportion must be maintained. Of this water, 70% is inside the body cells (*intracellular*) and the remaining 30% is in the body fluids (*extracellular*); 15–20% is in the *interstitial spaces* in the tissues, bathing all the body cells, even the bone cells, and the remaining 10–15% forms the fluid of the blood, i.e. the plasma and the lymph (Fig. 23.1). These three volumes of fluid are separated only by thin *semipermeable membranes*, the cell walls and the capillary walls; water constantly passes through these walls from one of these areas to the others, though the volume of each remains remarkably constant in normal health.

The water of which our bodies are largely made is, however, not static. Fresh water is taken in by the body each day and passes out of it by a number of channels. The total quantities taken in and passed out must balance one another. Water is taken in as water and other fluids drunk, and also in the foods eaten, which, like the body, consist largely of water. On average the healthy person takes in 1.5 litres of fluid as water and other drinks daily, and a little over 1 litre in his food, an average of 2600–2800 mL. A similar quantity is passed out of the body by the lungs as water vapour (400–500 mL), by the skin as sweat (500–600 mL), by the kidneys as urine (1000–1500 mL) and in the faeces, a small quantity (100–150 mL).

The quantities lost in urine, sweat and water vapour from the lungs vary with conditions. In hot weather and heavy work, more sweat is produced to cool the body, and less urine is passed; at the same time the loss of fluid causes thirst, so that more fluid is drunk. In fever the same changes take place; there is increased loss of fluid and there must be increased intake to balance it (Fig. 23.2).

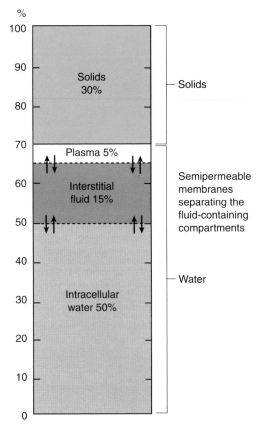

Fig. 23.1   Diagram to show the average percentage proportions by weight of solids and water, and the distribution of water in the body. The arrows indicate the passage of water and solutes between the three compartments containing them.

Electrolytes

In addition to the balance of the quantity of water in the body, the body fluids must be of the right composition, i.e. they must contain the correct balance of electrolytes. These *electrolytes* are minute particles split off from the molecules of the various salts and dissolved in water. They are called ions (see p. 24). They carry electrical charges and are of two types: negatively charged particles (anions) and positively charged particles (cations). The total of anions must balance the total of cations. The chief anions are chloride (Cl^-),

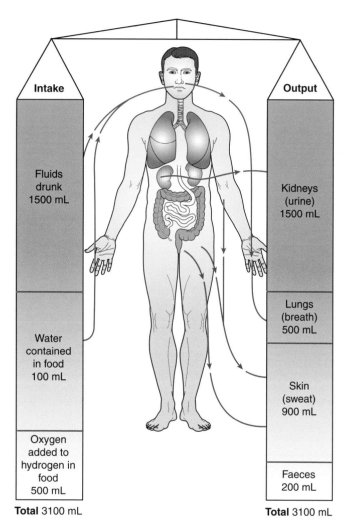

Fig. 23.2   Diagram to show how water balance is maintained by various organs.

bicarbonate (HCO_3^-) and phosphate (PO_4^{2-}); chloride and bicarbonate are present in quantity in plasma and interstitial fluid, while intracellular fluid contains mainly phosphate. The cations are sodium (Na^+) and potassium (K^+), with a little calcium (Ca^{2+}) and magnesium (Mg^{2+}). Sodium is present as the main positive ion in plasma and interstitial fluid, while intracellular fluid contains chiefly potassium.

ACID-BASE BALANCE

The acid–base balance of the body fluids affects their pH. Normally the body fluids are slightly alkaline and vary very little during life, since this reaction must be correct for the action of the various enzymes that control the cell activities, such as digestion, the use of food for growth and energy production, and the production and excretion of waste products. Normally, venous blood is a little less alkaline than arterial blood because of the carbon dioxide and other acids produced in the tissues. The interstitial and intracellular fluids are similar but slightly more alkaline. This balance is maintained by the 'buffers' these fluids contain. Alkalis and proteins neutralize acids produced by tissue activities, preventing acidosis, or ketosis, a condition that leads to coma and death if the body fluids become less alkaline than normal (the normal average is pH 7.4; see Chapter 1). In the same way, acids such as carbonate and chloride neutralize excess alkalis in the body fluids, preventing alkalosis, a condition that can prove fatal if it is not corrected. It may be caused by taking in an excess of alkaline salts such as sodium bicarbonate, or by loss of acid from excessive vomiting, or from continuous gastric aspiration after operation, or in respiratory conditions, or from the retention of normal quantities of an alkali such as potassium through renal failure.

THE URINARY SYSTEM

The urinary system consists of the following components (Fig. 23.3):

- The kidneys
- The ureters
- The bladder
- The urethra.

The kidneys

The kidneys are two bean-shaped organs (Fig. 23.4) situated in the posterior part of the abdomen, one on each side of the vertebral column, behind the peritoneum. They lie at the level of the twelfth thoracic to the third lumbar vertebrae, though the right kidney is usually slightly lower than the left because of its relationship to the liver.

Each kidney is about 11 cm long, 6 cm wide and 3 cm thick and is embedded in the perirenal fat.

The medial border of each kidney is concave in the centre. This area is called the *hilus* and it is the point at which the blood vessels, nerves and ureters enter or leave the kidney.

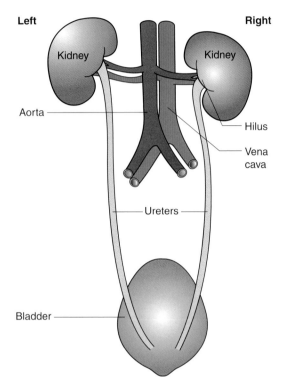

Fig. 23.3 The urinary system, viewed from the back.

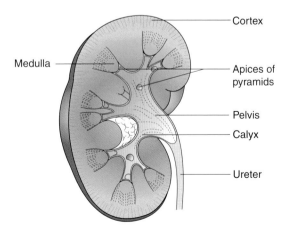

Fig. 23.4 Diagrammatic section of the kidney.

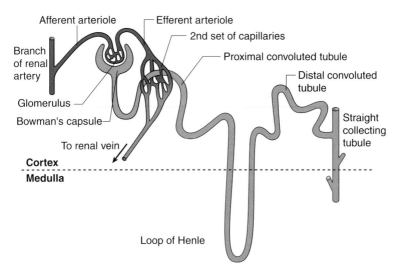

Afferent arteriole

Efferent arteriole

2nd set of capillaries

Branch
of renal
artery

Proximal convoluted tubule

Distal convoluted
tubule

Glomerulus

Bowman's capsule

Straight
collecting
tubule

To renal vein

Cortex

Medulla

Loop of Henle

Fig. 23.5   A nephron and the blood vessels associated with it.

The kidney is enclosed in a capsule of fibrous tissue, which can be easily stripped off. In vertical section the kidney has two distinct parts. The dark outer part is called the cortex and the paler inner portion the medulla. This leads into the collecting space, which is called the *renal pelvis*.

The kidney substance consists of minute twisted tubules called *nephrons* (Fig. 23.5); there are over a million in each kidney. Each of the nephrons begins in a cup-shaped expansion called the *glomerular capsule*, from which the tubule leads (Fig. 23.6). Into the cup of each capsule comes a fine branch of the renal artery, forming a tuft of capillaries in close contact with its inner wall; the capillary tuft is called the *glomerulus*. The arteriole bringing blood to the tuft is called the *afferent vessel*, and the arteriole that carries the blood away is the *efferent vessel*; it is slightly smaller than the afferent vessel. The blood in the tuft is under high pressure because of this and because of its nearness to the abdominal aorta.

The *convoluted tubule* makes a number of twists, the *proximal convolutions*, on leaving the capsule and then forms a long loop, the *loop of Henle*, which dips down into the medulla and passes back to the cortex. The tubule next makes a *distal* or second series of convolutions, and ultimately empties into a straight collecting tubule in the medulla.

The efferent blood vessel that comes from the capillary tuft or glomerulus in the capsule divides to form a second set of capillaries,

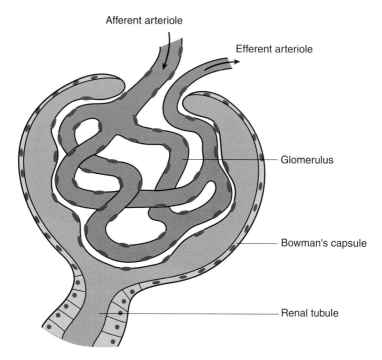

Fig. 23.6   Diagram of the glomerulus and its capsule.

around the walls of the convoluted tubules in the cortex. Thus the blood passes through two sets of capillaries within one organ, which does not happen in any other part of the body. The blood is collected from the second set of capillaries by small veins, which unite with other small veins to empty blood into the renal vein.

The production of urine

The function of the kidneys is to secrete and excrete urine. The composition of the blood must not vary beyond certain limits if the tissues are to remain healthy and this regulation depends on the removal of harmful waste products and the conservation of water and electrolytes in the body. *Urine* is produced by three processes:

1. *Filtration* under pressure occurs from the glomerulus where only the thin walls of the capillaries and of the glomerular capsule separate the blood from the kidney tubule. The walls of the glomerulus are permeable to water and other small molecules

but they are not permeable to blood cells or protein. Because the blood in the glomerulus is under pressure, some of the constituents pass through into the glomerular capsule. This fluid is known as the glomerular filtrate and it has a composition similar to plasma in that it contains glucose, amino acids, fatty acids, salts, urea and uric acid in the same proportions. Blood cells and protein molecules are only filtered if the kidney is diseased. About 600 mL of blood per minute pass through the glomerulus and of this about 125 mL become the glomerular filtrate. If this were all excreted, 150–180 L of urine would be passed each day. However, the average amount of urine passed each day is actually about 1.5 L so it is obvious that reabsorption must occur.

2. Selective *reabsorption* occurs because the cells lining the convoluted tubules are able to absorb the water, glucose, salts and their ions that the body needs. In normal health, all the glucose is reabsorbed and none is excreted in the urine. Most of the water and salts are also absorbed resulting in the 1.5 L of fluid that do pass into the collecting tubules containing normally about 2% urea. The acidity of the urine varies somewhat so that the reaction of the blood is maintained at a pH of about 7.4.

3. Active *secretion* occurs because the cells lining the tubule have the ability to secrete some substances from the blood in the second capillary network into the lumen of the tubule.

Reabsorption of water in the distal convoluted tubule is variable and is controlled by the secretion of *antidiuretic hormone* (ADH) from the posterior lobe of the pituitary gland. A decrease in secretion of ADH causes less water to be reabsorbed in the distal tubule, therefore more water is excreted as urine. The reabsorption of salts is controlled by the hormones of the adrenal cortex, especially *aldosterone*. The production of these hormones is increased or decreased according to the needs of the body to use water or salts and the electrolytes to which they give rise. Nervous control, together with the hormones epinephrine and norepinephrine, maintains the blood pressure at the high level required for filtration in the capillary tufts.

The medulla or inner portion of the kidney consists of straight collecting tubules into which the convoluted tubules of the cortex empty. It forms a number of cone-shaped masses, which project into the pelvis of the kidney. These are called the *pyramids* of the medulla and there can be from eight to twelve of them.

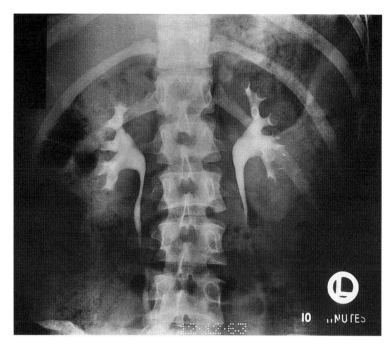

Fig. 23.7   A pyelogram showing the pelvis of the kidneys filled with an opaque drug, which has been excreted following an intravenous injection. Note that the right kidney is lower than the left. Note also the calyces outlining the apices of the pyramids of the medulla.

The apices of the pyramids project into the pelvis and are covered with the mouths of the fine collecting tubules, which pour the urine into the kidney pelvis. The function of the medulla is therefore to collect the urine secreted in the cortex and convey it to the pelvis.

The pelvis is an irregular branched cavity, which lies at the root or hilum of the kidney and leads like a funnel to the ureter (Fig. 23.7). Its branches, known as the *calyces* of the pelvis, penetrate into the kidney substance, and each branch receives the apex of one of the pyramids of the medulla. The pyramids pour the urine into the pelvis, which conveys it to the ureters.

The composition of urine

Normal urine, therefore, is formed partly by filtration under pressure from the capsules and partly by reabsorption and by secretion in the tubules. Urine is an amber fluid varying in colour according to its quantity. It is an acid and has a specific gravity of 1015 to 1025 (for a definition of specific gravity, see p. 12).

Urine consists of water, salts and protein waste products, namely urea, uric acid and creatinine. The average composition is water (96%), urea (2%), and uric acid and salts (2%).

The percentage of urea in blood plasma is 0.04 compared with 2 in the urine; hence the concentration has been increased 50 times by the work of the kidney. The salts consist chiefly of sodium chloride (NaCl), with phosphates and sulphates, produced partly from the use of the phosphorus and sulphur present in protein foods. These salts must be reabsorbed or eliminated in the quantities necessary to keep the blood at its normal composition. Since this composition is essential to the life of the blood corpuscles and tissue cells, this function of the kidney is very important. The normal quantity of urine secreted is 1.5 L in 24 hours but it is increased by drinking and by cold weather, and is decreased by reducing fluid intake and by hot weather, exercise, and fever, since these increase sweating. Potassium salts are normally filtered out, and are reabsorbed or excreted as required, to keep the correct level in the body fluids. In renal failure, their excretion may be checked so that the amount in the body fluids and tissues rises.

Countercurrent exchange

A unique mechanism is thought to operate in the kidney to ensure that, when it is required, a concentrated urine can be produced to conserve water. This mechanism is called *countercurrent exchange* (Fig. 23.8). By this mechanism, NaCl is transported out of the ascending loop of Henle and enters the descending loop. This leads to a *concentration gradient* of NaCl in the kidney, with the highest concentration being at the base of the loop of Henle. A similar mechanism operates in the surrounding blood vessels to ensure that, by the process of osmosis (see Chapter 1), the water drawn into the tissue surrounding the loop is taken away in the blood stream.

The outcome of this mechanism is that the collecting tube passes through a concentration gradient of NaCl. When water does not need to be conserved the collecting tube is impermeable to it and a dilute urine is produced. However, when water requires to be conserved, ADH acts to make the collecting tube permeable and water is drawn into the kidneys by osmosis. The water is taken away into the blood stream by the mechanism outlined above.

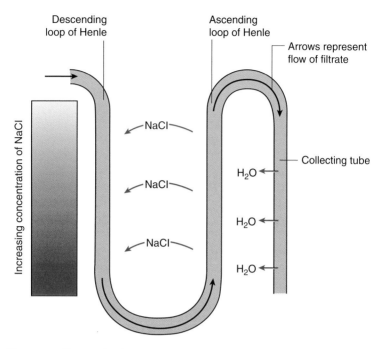

Fig. 23.8   The mechanism of countercurrent exchange.

The ureters

The ureters are the two tubes that carry the urine from the kidneys to the bladder. Each is about 25–30 cm long and is a thick-walled, narrow tube continuous with the renal pelvis and opening into the base of the bladder. It is about 3 mm in diameter but is slightly constricted in three places: at the junction with the renal pelvis; where it crosses the brim of the lesser pelvis; and as it passes through the wall of the bladder. These narrowed portions may be the site of impaction of a *ureteric calculus* (stone). The ureters, the renal pelvis and the calyces can be seen by radiography following the intravenous injection of a radiopaque substance.

The ureter has an outer fibrous coat continuous with the fibrous capsule of the kidney, a muscular coat with an outer circular and an inner longitudinal layer, and a lining of mucous membrane, which is continuous with the lining of the bladder. The muscular layer of the ureter undergoes peristaltic contractions, usually about four or five times a minute.

The bladder

The bladder is a reservoir for urine and its size, shape and position vary with the amount of fluid it contains. When empty, it lies within the lesser pelvis but as it becomes distended with urine it expands upwards and forwards into the abdominal cavity.

Both ureters enter and the urethra leaves the bladder at its base. An imaginary line drawn to connect these openings outlines the area known as the trigone. The neck of the bladder is the lowest and most fixed part of the organ; it lies 3–4 cm behind the symphysis pubis. The bladder can hold over 500 mL of urine though this would be painful; the desire to empty the bladder is normally felt when it contains around 250–300 mL of urine.

The bladder has three coats. The outer *serous layer* is of peritoneum but this is found only on the superior surface. The *muscular layer* contains both circular and longitudinal muscle fibres. There are also two bands of oblique fibres, which are situated close to the ureteric openings and which prevent urine flowing back into the ureters. The *inner mucous coat* is loose and is thrown into rugae when the bladder is empty. The bladder is lined with transitional epithelial tissue, which allows for expansion when the organ is full.

The urethra

The urethra extends from the internal urethral orifice in the bladder to the external urethral orifice.

In men the urethra is 18–20 cm long and serves as a common canal for both reproductive and urinary systems. It is divided into three portions:

- The prostatic portion is about 3 cm long and is surrounded by the prostate gland. It is lined with transitional epithelium, and the orifices of the prostatic ducts and the ejaculatory ducts open into it.
- The membranous portion is 1–2 cm long and is the narrowest part of the urethra. It passes through the pelvic floor.
- The spongy portion is about 15 cm long and lies within the penis.

In women the urethra is about 4 cm long and serves the urinary system only. It begins at the internal urethral orifice of the bladder and passes downwards behind the symphysis pubis embedded in the anterior wall of the vagina.

There are internal and external sphincters to the urethra; the internal sphincter is involuntary and the external sphincter is under voluntary control except in early infancy and in nerve injury or disease.

Micturition

Micturition is the passing of urine. Urine is constantly passing into the bladder from the ureters; when there is 200–300 mL of urine in the bladder the desire to pass urine occurs due to stimulation of the sensory nerves because of increased tension in the bladder. As the sensory impulses increase in number and frequency the motor impulses cause a reflex contraction of the bladder and relaxation of the internal sphincter. The external sphincter is controlled by the *pudendal nerve*. When a child has learned to inhibit spinal reflexes, micturition can be delayed for a considerable time or can be induced voluntarily.

In disease, micturition may be affected in many ways. The sphincter may be paralysed in a state of contraction (spastic paralysis) so that it cannot be relaxed. This will cause retention of the urine and the bladder will become overfull. If this is not relieved by catheterization, distension will continue and will open the orifice that the sphincter guards, allowing a little urine to dribble away continuously though the bladder still remains full. This is called retention with overflow and is bad for both bladder and kidney, causing loss of tone and interference with blood supply to the stretched bladder wall and back pressure on the kidney. It should never be allowed to occur.

The sphincter may be paralysed in a state of relaxation. In such conditions, urine will dribble constantly from an empty bladder, as the bladder cannot retain it. This is comparatively rare.

In other cases of nervous disease and in anaesthesia the control of the brain over micturition may be abolished, so that the act again becomes reflex as in the lower animals – the bladder fills and empties by reflex, without the individual being aware of any sensation of fullness or being able to control the relaxation of the sphincter.

SELF-TEST QUESTIONS

- List the typical input and output of fluids.
- Describe glomerular function.
- What are the stages of urine production?
- What is the function of the countercurrent mechanism in the kidney?

SECTION 6

Protection and reproduction

SECTION INTRODUCTION

Protection and reproduction are considered in the same section
because they are vital for the immediate survival of individuals and
the continued survival of the human race. The first line of defence,
the skin, is mainly physical and it protects us by preventing
dehydration and the entry of harmful foreign bodies such as bacteria.
The skin also has cellular protective mechanisms and these are
evident in the process of inflammation.

Protection at the cellular level takes place non-specifically, i.e. in
response to invasion by any foreign body, and specifically by the
development of immunity. Several cell types, which are located in a
number of places in the body including the blood, the lymphatics and
the skin, are involved and there is an intimate relationship between
the non-specific defences and immunity.

Reproduction is possible due, essentially, to the production of
different types of gametes: sperm and ova in males and females,
respectively. The respective reproduction systems of males and
females differ anatomically and in their physiological regulation but
both are ultimately designed to bring together sperm and ova,
through sexual intercourse, in order to achieve fertilization.

Chapter 24

The skin

CHAPTER CONTENTS

LEARNING OBJECTIVES

After reading this chapter you should understand:

- the structure of skin
- the function of skin
- the repair of skin

This chapter covers the gross structure of the skin and the associated structures such as sweat glands and sebaceous glands. Two functions of the skin, namely temperature regulation and wound healing, are considered.

The skin covers the body and protects the deeper tissues. It contains the endings of many sensory nerves and is important in the regulation of body temperature.

STRUCTURE OF THE SKIN

The skin has two layers:

- The *epidermis* or *outer layer*
- The *dermis*.

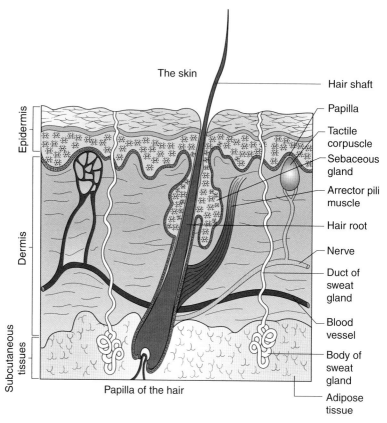

Fig. 24.1   Diagrammatic section of the skin.

The *epidermis* is non-vascular and consists of *stratified epithelium* (Fig. 24.1). It is very thick, hard and horny on such areas as the palms of the hands and soles of the feet and is much thinner and softer over other parts, such as the trunk and the inner sides of the limbs. The epidermis has two layers or zones: the outer is the horny zone; the inner is the germinative zone (Fig. 24.2). The horny zone has three layers:

- The *horny layer* (stratum corneum) is the most superficial layer; the cells are flat and have no nuclei and the protoplasm has been changed into a horny substance called *keratin*, which is waterproof

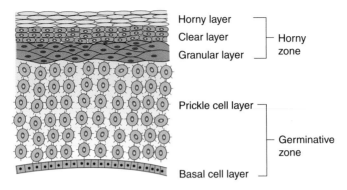

Fig. 24.2 The epidermis.

- The *clear layer* (stratum lucidum) is composed of cells with clear protoplasm, and some cells have flattened nuclei
- The *granular layer* (stratum granulosum) is the deepest layer and consists of several layers of cells with granular protoplasm and distinct nuclei.

The *germinative zone*, which is deeper, consists of two layers:

- The prickle cell layer contains cells of varying shapes, each having short processes, rather like thorns, joining them together; the nuclei are distinct
- The basal cell layer consists of columnar cells arranged on a basement membrane.

The surface scales are constantly being rubbed off by friction and are constantly renewed from below as the deep cells multiply and are driven up to the surface, developing into scales as they approach it. The epidermis has no blood supply and practically no nerve supply. It is nourished by lymph from the blood vessels in the underlying dermis. It is the epidermis that is raised in a blister, and the blister can therefore be snipped with scissors without causing any pain, to allow the lymph it contains to escape.

The *corium* is a tough elastic layer, which is very thick, in the palms of the hands and soles of the feet and very thin in the eyelids. It consists of connective tissue with elastic fibres, blood vessels, lymphatic vessels and nerves. Numerous conical projections, called papillae, extend from the surface of the corium and protrude into the epidermis. They are most numerous where the skin is most sensitive and in these areas they are arranged in parallel ridges, different in each individual, which are useful as fingerprints.

The basal layer of cells in contact with the *corium* contains the pigments, which give the skin its colour. The pigment protects the body from the harmful effects of the sun's rays as dark colours absorb radiation. The scales of the surface prevent the entrance of bacteria into the tissues since they cannot digest these dried-up cells and make their way through them. Once the epidermis is broken by a cut or prick, infection may enter the tissues and sepsis may follow.

The nerve endings in the skin are for the most part sensory, and are of different varieties to give the various different sensations of which the skin is capable, namely, the sensations of touch, heat, cold and pain. The nerves of touch end in round bodies known as the touch or tactile corpuscles, which are stimulated by pressure, and the nerves of heat, cold and pain in delicate, tree-like branches. A few of the branches of these nerve endings pass into the epidermis. Heat is experienced only if a hot object touches the skin over the termination of a special nerve ending affected by heat. In some parts the nerve endings are so close together that this is not appreciated but where the nerve endings are less numerous, as over the back of the hand, it is possible to find spots where heat can be felt (hot spots) and others where cold is felt (cold spots).

The arteries supplying the skin form a network in the subcutaneous tissue and branches supply the sweat glands and the hair follicles. Fine capillaries also pass into the papillae.

Appendages of the skin

The skin carries four appendages:

- Sweat glands
- Hairs
- Nails
- Sebaceous glands.

The *sweat glands* are twisted, tubular glands that lie deep in the true skin. Their ducts open in the pores of the epidermis, and the tube is coiled up into a little round ball deep in the skin, called the body of the gland. The sweat glands secrete *sweat*, which consists of water, salts and a trace of other waste products. Much of this sweat evaporates immediately it reaches the skin surface, and is called *insensible sweat*. When sweating is heavy, some of the sweat is poured out on to the skin, which becomes wet; this is called *sensible sweat*, and evaporation takes place from the whole surface. If sweating is profuse and runs off the body, the cooling effect is lost. The secretion of sweat is a means of excretion of waste products, and some poisons and drugs may be eliminated in this manner. Its chief importance, however, lies in the fact that the evaporation of sweat uses up body heat, since heat is required to turn water into water vapour. The amount of sweat secreted therefore depends on the amount of heat that the body needs to lose. The average amount excreted in 24 hours is 500–600 mL. In hot weather and in violent exercise, sweating is heavy, so that much heat is lost by the evaporation of sweat. At these times, less urine is passed so that there is not an excessive loss of fluid. In cold weather or rest, less sweat is secreted so that there is less loss of heat by evaporation. At the same time more urine is secreted by the kidney so that loss of water is balanced.

Sweat glands are present all over the body but are larger and more numerous in certain parts, such as the palm of the hand, the sole of the foot, the axillae, groins and forehead.

The *hairs* consist of modified epithelium. They grow from tiny pits in the skin, known as *hair follicles*. At the base of each follicle is a group of epithelial cells forming the root from which the hair grows. The hair root is the part of the hair within the follicle, extending through both dermis and epidermis. The hair shaft projects beyond the epidermis. The *hair bulb* is the expanded part of the hair within the follicle. At its base is a small conical projection, called a *papilla*, which contains blood vessels and nerves to supply the hair. Hairs are always set obliquely in the skin. The *arrectores pilorum* are small involuntary muscles connected to the hair follicles. They are always on the side toward which the hair leans so that when they contract the hair stands erect. The skin around the hair is elevated at the same time, which produces the effect known as 'goose flesh'. The hairs are constantly being shed and renewed. As long as the root is healthy fresh hair will grow from it but if the root is destroyed or its blood supply interfered with, the growth of the hair will stop. On the scalp, baldness results. Hair is present all over the body except on the palms of the hands and soles of the feet but is so fine and sparse as a rule that it cannot be seen. It is longer and more plentiful in the eyebrows, axillae and groins, where it entangles the sweat, preventing it from running down and assisting its evaporation.

The *nails* are horny plates of modified epithelium that protect the tips of the digits. They grow from a root of typical soft epithelial cells at the base of the nail. This root is embedded in the fold of the epidermis. The nails replace the epidermis here but are continuous with it so that there is a continuous barrier to exclude bacteria. It is, however, easy for a break in the continuity to occur here if the nails are not carefully looked after and this is particularly dangerous among nurses, since they are in contact with infection in the course of their work. In animals the nails are protective in more senses than one but as people have invented implements of various kinds, human nails are not usually worn down as in animals and so need regular cutting.

The *sebaceous glands* are small saccular glands that secrete an oily substance called *sebum*. They are situated in the angle between the hair follicle and the arrector pili muscle so that contraction of the muscle has the effect of squeezing sebum from the gland. This lubricates the skin and hair, and keeps them soft and pliant, so that they do not readily break. However, sebum picks up dust and bacteria, which cling to an oily surface. As a result we must constantly remove it by washing with soap and water. If some substitute is not applied the skin quickly becomes brittle so that it breaks readily and lets bacteria enter. The nurse, whose hands are often in lotions and soapy water, which remove the sebum, needs to pay particular attention to this.

FUNCTIONS OF THE SKIN

The functions of the skin are:

- To regulate body temperature
- To secrete waste products
- To make us aware of our environment, as the organ of touch and other senses
- To keep out bacteria by its dry, scaly outer surface
- To secrete sebum
- To protect the body by its pigment from the harmful effects of the sun's rays
- To produce vitamin D through the action of UV rays on the ergosterol it contains.

Control of body temperature

Body temperature is a balance between heat gained and heat lost. Human beings are warm-blooded animals and the temperature must be maintained around 37°C; a rise or fall of 1°C or more affects the normal functioning of the nervous system and enzymes.

The main temperature-regulating mechanism is in the hypothalamus. It acts by negative feedback: if the body temperature rises, mechanisms come into action so that heat is lost from the body; if the body temperature falls, heat is conserved until the temperature approaches normal.

Heat production occurs primarily by metabolic activity. Additional heat is produced by exercise, activity, increased muscle tension and shivering, and also by endocrine disorders, infections, trauma and emotion. Heat production is lowest during sleep and highest during muscular activity. Heat loss occurs through the following mechanisms:

- Radiation, conduction and convection of heat from the skin
- Evaporation of sweat
- Respiration
- Excretion of urine and faeces.

Radiation is the transfer of heat from one object to another without physical contact between the two. The body radiates heat to every object near it and the heat loss is proportional to the surface area; a large area loses a lot of heat but heat loss can be diminished by reducing the surface area. For example, less heat is lost from a body in a curled-up position than in a stretched-out position.

Conduction is the transfer of heat from one molecule to another. If one end of a metal rod is placed in the fire, heat will be conducted along it until the whole rod is hot. Direct contact will cause the body to lose heat to any object that is cooler than the body.

Convection is the transfer of heat from the body to the air, which then rises and is replaced by cooler air, which in turn is heated. Heat loss by convection is reduced by wearing suitable clothing.

Evaporation of sweat from the surface of the skin is continuous and has a cooling effect on the body. It is more effective in dry climates because when the air is humid and already saturated with water vapour further evaporation cannot occur.

Heat is also lost from the body with every exhalation of air since expired air contains water vapour, which evaporates. Only small amounts of body heat are lost through the passage of urine and faeces.

In hot weather, to keep the temperature normal:

1. Heat production is lessened; the thyroid and adrenal glands do not stimulate so much tissue activity.

2. Heat loss is increased by dilation of blood vessels in the skin so that radiation, conduction and convection are increased and there is increased sweating, so more heat is lost by evaporation.

Ageing and the skin

As the skin ages it becomes thinner (atrophied) and the blood supply decreases. In older people who are ill, especially if they are dehydrated, pyrexial or malnourished, there is an increased risk of the skin breaking down, especially if it is subjected to pressure or shearing forces. A break in the skin caused by this is know as a pressure ulcer.

Wound healing

When skin is damaged, for instance by a cut, it has the ability to repair itself by the process of wound healing. Superficial abrasions of the skin heal by regrowth of the epidermis to cover the damaged area.

If the damage extends to the blood vessels, the prevention of blood loss precedes the healing of the wound, and this takes place by haemostasis involving clotting of the blood (see Chapter 14). The clot is dissolved by the action of plasmin below the surface of the wound. This allows the healing process to take place.

The process of healing (Fig. 24.3) begins with a non-specific *inflammatory response* (see Chapter 25). The inflammation is required to bring white blood cells to the site of injury, where they remove foreign material, such as bacteria, which may lead to infection.

The next stage of healing is called the *migratory phase*, during which epithelial cells migrate beneath the clot, thereby forming a scab. Fibroblast cells, which are responsible for producing collagen, also migrate into the wound. At the same time the damaged blood vessels repair and grow in the granulation phase.

The *proliferative phase* is characterized by the continued growth of epithelium beneath the scab, the continued growth of blood vessels and the production of fibres of collagen by the fibroblasts. The collagen gives strength to the healing wound.

Finally, in the *maturation phase*, which may take many months depending upon the extent of the wound, collagen fibres become

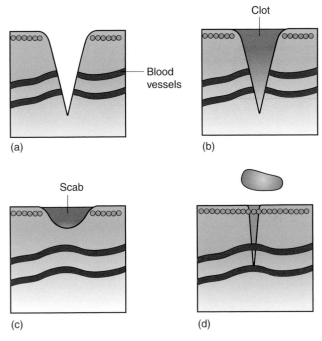

Fig. 24.3 Wound healing. (a) Injury; (b) clotting; (c) migration, granulation, proliferation; (d) maturation.

more organized and pull the edges of the wound together. The number of fibroblasts decreases and a normal blood supply is restored.

SELF-TEST QUESTIONS

- What are the two layers of the skin?
- What are the appendages of the skin?
- List the functions of the skin.
- How is heat lost from the body?
- Describe the stages of wound healing.

Chapter 25

Non-specific resistance and the immune system

LEARNING OBJECTIVES

After reading this chapter you should understand:

- the differences between specific and non-specific protection
- the process of inflammation
- the types of immunity
- the monocyte–macrophage system

If the physical barriers of the body, such as the skin and mucous membranes are penetrated, then the body is still able to defend itself against potentially harmful material. These harmful materials, called foreign bodies because they are not recognized by the body as being part of itself, include bacteria, viruses and tissue from other organisms, e.g. transplants.

There are two ways in which the body defends itself: one is non-specific and will be activated whenever foreign bodies invade and one is specific – the immune system – and responds to specific foreign bodies. The non-specific defences are considered first.

INFLAMMATION

If the skin is cut (see Chapter 24) then chemicals are released from the damaged tissue; these initiate an *inflammatory response* (Fig. 25.1). Inflammation can be identified by four cardinal signs:

- *Redness*
- *Pain*
- *Heat*
- *Swelling*

and sometimes *loss of function*.

The outcome of the inflammatory response is that *vasodilation* (swelling of blood vessels) takes place. This increases their permeability, and *macrophagocyte cells* migrate to the site of the injury. The macrophagocyte cells include monocytes and neutrophils and these are responsible for ingesting foreign bodies and destroying them. The inflammatory response is also necessary for the initiation of wound healing (see Chapter 24). The basophil cells, which are called *mast cells* when they are in the skin, play a part in this inflammatory response by releasing *histamine*, which leads to inflammation.

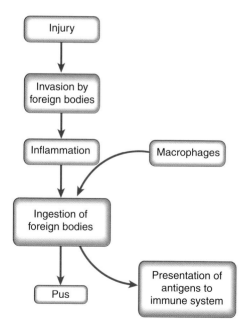

Fig. 25.1 Non-specific defences.

A collection of white cells at a site of injury can be recognized as the white material *pus*, which is a collection of dead cells and fluid. In some cases of injury and inflammation the body temperature can rise for prolonged periods above normal body temperature, and this is called *pyrexia*. While pyrexia can be very uncomfortable and even dangerous if too prolonged, it is a response of the body aimed at speeding up all the processes that deal with invasion by foreign matter.

Cells of the non-specific defences, along with T-lymphocytes, play a role in the immune system as *antigen-presenting cells* (APCs). By ingesting foreign bodies the APCs subsequently display parts of these foreign bodies or antigens on their surface, at the outside of the plasma membrane, and these are part of the process that triggers the immune system.

THE IMMUNE SYSTEM

There are two types of immunity:

- Cell-mediated immunity
- Antibody-mediated immunity.

The T- and B-lymphocytes, respectively, are responsible for *cell-mediated* and *antibody-mediated immunity* (Fig. 25.2). In response to an antigen, the lymphocytes produce T-cells, which can attach to and ingest the foreign bodies, and also B-cells, which produce specific protein molecules called *antibodies*, which can also attach to *antigens*. By attaching to the antigens, they are inactivated and the resulting *immune complexes* are later ingested by the eosinophils.

Three types of T-lymphocyte are produced: the *T-killer cell*, which actively ingests foreign bodies; the *T-helper cell*, which helps to activate the production of antibodies; and the *T-memory cell*, which is responsible for subsequent specific recognition of invasion by the antigen. Together with the B-memory cell, the T-memory cell is the basis of immunity.

IMMUNITY

When a foreign body is introduced into the body the immediate response is the production of a substance that will react with it and render it harmless. The foreign protein is called an *antigen* and substances produced in response to the antigen are called *antibodies*. Antigens can be any foreign protein although common antigens are micro-organisms, some drugs (such as penicillin), animal and

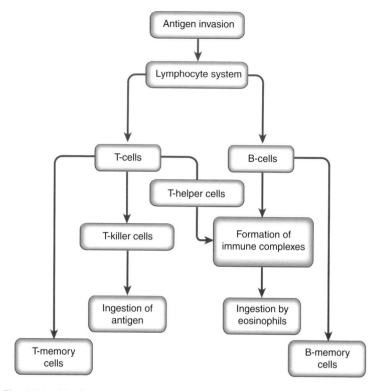

Fig. 25.2 The immune system.

vegetable proteins (including pollens) and foreign tissues (such as transplanted organs). The reaction that occurs may be called an antigen–antibody reaction. When it occurs as a response to micro-organisms it is called immunity.

Immunity is a useful defence against infection. Each type of micro-organism which enters the body acts as an antigen and stimulates the production of specific antibody that destroys that antigen and no other. The first time a person comes into contact with the virus that causes measles, an antibody is produced which enables the body to overcome the infection and which remains in the blood ready to prevent any further infection by the same type of virus. Immunity may be *active*, when the cells of the body make the antibody, or may be *passive*, when the antibody has been produced in the cells of another person (Fig. 25.3).

Active immunity may be achieved in several ways. *Active natural immunity* is gained by having the disease, after which the antibody

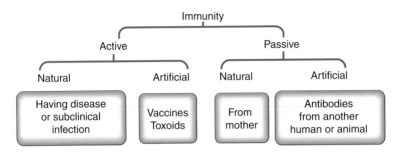

Fig. 25.3   Active and passive immunity.

remains in the blood ready to prevent another attack of the same disease. This type of immunity is also developed by subclinical infections, in which the body is exposed to small numbers of micro-organisms insufficient in number to give rise to any definite symptoms but which are sufficient to stimulate antibody production. *Active artificial immunity* is given to children and travellers to prevent their developing diseases that would be serious or fatal. An injection of killed micro-organisms, or live ones that have been rendered harmless, is given and the body responds by producing antibodies. In this way, active immunity is built up. Harmless toxins are also used to give this type of immunity. *Toxins* are chemical poisons produced by micro-organisms and, when rendered harmless, they also act as antigens. Harmless micro-organisms are called *vaccines* and harmless toxins are called *toxoids*. Many disease are prevented by active artificial immunity; some of the most common are whooping cough, diphtheria, measles, smallpox, poliomyelitis and tuberculosis.

Passive natural immunity is gained by the baby before birth as the antibodies are passed from the mother to the fetus. *Passive artificial immunity* is useful in the prevention of disease or in its treatment. The antibodies are produced in another human or in an animal and are injected into the person at risk. Passive immunity is always short-lived as the antibodies are destroyed after a short time.

Antigen–antibody reactions normally occur in the blood stream and the debris is carried away by the *monocyte–macrophage* system. When the immune reaction occurs in the tissues themselves the cells are damaged or destroyed by its side-effects and this is known as allergy. Allergic reactions are often due to protein-like substances called allergens. The allergic reaction in the tissues releases histamine, which causes redness and swelling in the skin, as in *urticaria*, and the outpouring of fluid as in *hay fever*. There may also be constriction of smooth muscle in the respiratory tract causing *asthma*.

The word *autoimmunity* describes the situation in which the body makes antibodies against some of its own cells. Many diseases are thought to be autoimmune in origin, two of the best known being *rheumatoid arthritis* and *rheumatic fever*.

Monocyte–macrophage system

Monocytes have already been described as being one type of white blood cell. They are actively phagocytic and mention was made of their being part of the monocyte–macrophage system. This system is widespread throughout the body and its cells have the ability to move about which, combined with their phagocytic powers, makes them one of the most important defences of the body against micro-organisms. They are also concerned with the production of anti-bodies. The cells of the system occur in five parts of the body:

- Connective tissue, where they are called histiocytes
- Blood, where they are called monocytes
- Lining blood vessels of the bone marrow, spleen and liver and parts of the adrenal gland and the anterior lobe of the pituitary
- In the lymph nodes, lymph follicles of the small intestine and the tonsils
- The meninges, where they are called meningocytes.

SELF-TEST QUESTIONS

- Describe the stages of inflammation.
- What do macrophages do?
- Describe cell-mediated immunity.
- Contrast active and passive immunity.
- Contrast natural and artificial immunity.
- Where are the cells of the monocyte–macrophage system?

Chapter 26

The reproductive system

LEARNING OBJECTIVES

After reading this chapter you should understand:

- the male reproductive system
- the female reproductive system
- the menstrual cycle

THE MALE GENITAL ORGANS

The male genital organs (Fig. 26.1) consist of six components:

- The testes and epididymides
- The deferent ducts
- The seminal vesicles
- The ejaculatory ducts and penis
- The prostate
- The bulbourethral glands.

The *testes* (Fig. 26.2) are the reproductive glands in the male. They are suspended in the *scrotum* by the *spermatic cords* but they develop high up in the abdomen close to the kidneys and gradually descend through the inguinal canal into the scrotum shortly before birth.

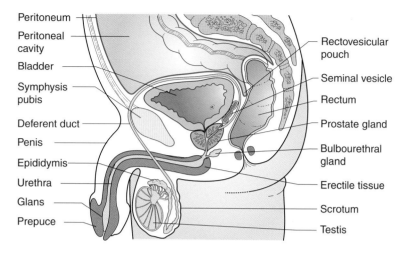

Fig. 26.1   The male reproductive organs.

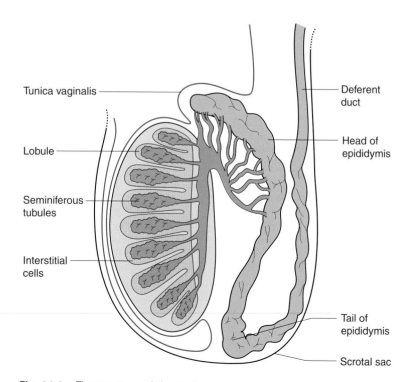

Fig. 26.2   The structure of the testis.

Occasionally one, or both, glands fails to descend and remains in the abdomen or in the inguinal canal; surgery may then be required to relocate it (or them).

As the testis descends, it brings down with it a pouch of peritoneum called the *tunica vaginalis*. This forms a serous covering for the testis but the pouch should otherwise be obliterated. If it is not obliterated, it forms a possible site for *hernia*. This causes an *inguinal hernia*, a loop of bowel or some other organ falling into the pouch, which forms the hernia sac. Each testis consists of 200 to 300 lobules, each containing up to three tiny convoluted tubules called the convoluted *seminiferous tubules*. The epithelial lining of their walls contains cells that develop into *spermatozoa* by a process of cell division. The tubules are supported by loose connective tissue, which contains groups of *interstitial cells*; these cells secrete the male hormone testosterone.

The *epididymis* is a fine, tightly coiled tube packed into the form of a long narrow body, which is attached to the back of the testis. The *seminiferous tubules* of the testis open into it and it leads into the deferent duct.

The *deferent duct* is a continuation of the duct of the epididymis. It passes through the inguinal canal and runs between the base of the bladder and the rectum to the base of the *prostate gland* where it is joined by the duct of the seminal vesicle.

The *seminal vesicles* (Fig. 26.3) are two pouches lying between the base of the bladder and the rectum. They secrete an alkaline fluid containing nourishment, which forms a large part of the seminal fluid.

The *ejaculatory ducts* are formed by the union of the ducts of the seminal vesicles and the deferent ducts. They commence at the base of the prostate and end at the opening of the prostatic utricle in the urethra.

The *penis* is a tubular organ plentifully supplied with large venous sinuses that can fill with blood causing *erection* of the organ. It contains the urethra, which is common to both the urinary and reproductive systems in the male. At the tip of the penis is an enlargement called the *glans penis*, in the centre of which lies the urinary meatus. The glans is normally covered by a loose double fold of skin called the prepuce or *foreskin*. It should be possible to draw the foreskin back over the glans penis but sometimes the opening in it is too small. This is known as *phimosis* and is treated either by stretching the foreskin or by circumcision, i.e. cutting it away, in more severe cases.

The *prostate* surrounds the commencement of the urethra in the male. It is about the size of a chestnut and contains the urethra and the ejaculatory ducts. It consists partly of glandular tissue and partly of involuntary muscle and produces a secretion called *semen*, which is alkaline and provides nourishment for the sperm.

The *bulbourethral glands* are situated on either side of the membranous portion of the urethra. The ducts open into the spongy portion of the urethra and the glands secrete a substance that forms part of the seminal fluid.

The *seminal fluid* is composed of substances secreted by the testes, the seminal vesicles and the prostate; it contains the spermatozoa.

The spermatozoa are minute cells each with a tail-like projection joined to the cell by a constricted portion called the neck (Fig. 26.4).

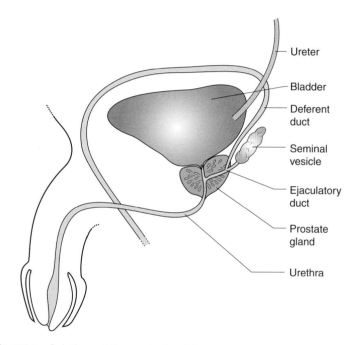

Ureter

Bladder

Deferent duct

Seminal vesicle

Ejaculatory duct

Prostate gland

Urethra

Fig. 26.3   Relations of the seminal vesicles.

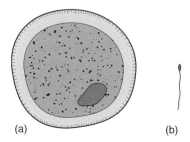

(a)   (b)

Fig. 26.4   (a) A mature ovum. (b) A spermatozoon, drawn to the same scale as (a).

The tail has a lashing movement that enables the cell to move after the semen leaves the male reproductive tract. As a result, when the spermatozoa are deposited in the vagina they can make their way up the uterus and uterine tubes in search of the ova. They are produced in enormous numbers and it is estimated that on average, 300 000 000 are deposited in the vagina at one time, though only one is necessary to fertilize the ovum.

THE FEMALE GENITAL ORGANS

The female genital organs consist of an internal and an external group (Fig. 26.5).

Internal organs

The internal organs, situated within the lesser pelvis, are:

- The ovaries
- The uterine tubes
- The uterus
- The vagina.

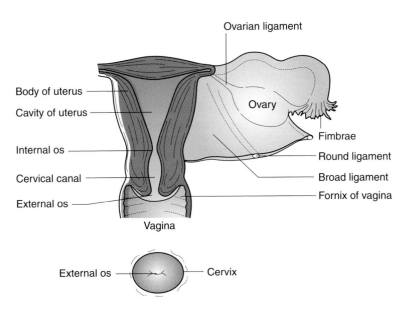

Fig. 26.5 The female reproductive organs, from behind.

The *ovaries* (Fig. 26.6) are two small glands about the size and shape of almonds, which are situated in the lesser pelvis, one on either side of the *uterus*, behind and below the *uterine tubes*. Each is attached to the broad ligament by a fold called the *mesovarium*. The fimbriated ends of the uterine tube and a *suspensory ligament* are also attached to the *ovary*.

After *puberty* the ovary has a thick cortex surrounding a very vascular medulla. At birth the cortex contains numerous primary *ovarian follicles*. After puberty, some develop each month to form vesicular ovarian follicles (graafian follicles) one of which usually matures and ruptures, releasing an *ovum*. This process is called ovulation. The ovum passes into the uterine tube along its fimbriated end and may be fertilized by a male sperm. If fertilization occurs, it usually takes place in the lateral third of the uterine tube.

After ovulation the vesicular ovarian follicle is converted to a mass of specialized tissue called the *corpus luteum*. If fertilization occurs the corpus luteum remains active until late in the pregnancy; if fertilization does not occur the corpus luteum begins to degenerate after about 14 days. The corpus luteum produces the hormones *progesterone* and *oestrogen*; these hormones cause the lining of the uterus, the *endometrium*, to become thickened, ready to receive the fertilized ovum. However, if fertilization does not occur, the hormones are withdrawn as the corpus luteum degenerates and the endometrium is shed in the process called menstruation (see p. 371).

The functions of the ovary are controlled by hormones from the pituitary: follicle-stimulating hormone (FSH) and luteinizing hormone.

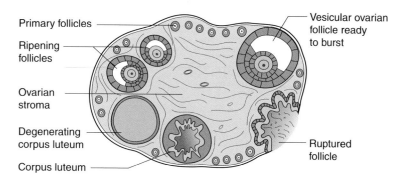

Fig. 26.6 The ovary, showing primary follicles, ripening follicles and corpus luteum.

The ovaries begin to function at puberty and continue to discharge ova at monthly intervals from about the age of 13 to about the age of 45 years, the usual time of the *menopause*. The ovaries are smooth in childhood but the scars that follow the rupture of the follicles pucker up the surface, so that they finally become very irregular and rather like almonds in appearance as well as shape.

Hormones from the ovaries are also responsible for the development of the reproductive system and the general development that marks *puberty* in the female, and occurs at about 13 years of age. There is marked development of the external genitals, of the uterus and the breasts: hair grows on the genitals and in the axillae; there is a general rounding of the figure and a gradual development of the typical mental outlook of the female personality, which gradually matures as the reproductive organs themselves mature.

The uterine tubes are situated in the upper part of the broad ligaments of the uterus. They are about 10 cm long and transmit the ova from the ovaries to the cavity of the uterus. Each tube has four parts:

1. The infundibulum is a trumpet-shaped expansion that opens into the abdominal cavity close to the ovary and has a number of processes called fimbriae.
2. The ampulla is a thin-walled, tortuous part that forms rather more than half the tube.
3. The isthmus is round and forms about one-third of the tube.
4. The uterine part passes through the wall of the uterus and is about 1 cm in length.

The uterine tubes have three coats: an outer serous covering of peritoneum; a muscular coat; and a lining of ciliated epithelium. The ova are conveyed along the tube by the peristaltic action of the muscular coat and by the action of the cilia.

The *uterus* is a hollow, thick-walled, muscular organ situated in the lesser pelvis between the rectum and bladder (Fig. 26.7). It is about 7.5 cm long, 5 cm across and 2.5 cm thick and it weighs about 30 g. It communicates with the uterine tubes, which open into the upper part of the uterus, and the *vagina*, which leads from the lower part. The uterus forms almost a right angle with the vagina, into which the *cervix* of the uterus protrudes. The upper part of the uterus is broad and is called the *body of the uterus*; the part of the body above the entrance of the uterine tubes is called the *fundus*. The cervix is narrower and more cylindrical than the body and projects through the anterior vaginal wall. The narrow opening at the upper end is called the *internal os* while the lower opening is called the *external os*.

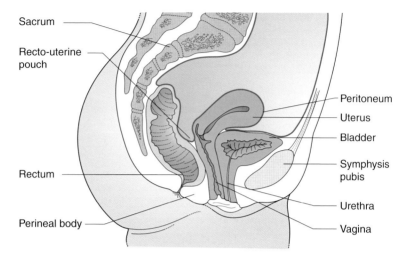

Fig. 26.7   Median section of the female pelvis, showing the position of the uterus.

The uterus has three coats: the outer serous coat is derived from the peritoneum; the muscular coat consists of involuntary muscle, the fibres of which are arranged in longitudinal, oblique, transverse and circular layers; and the lining is of mucous membrane and columnar epithelium known as the *endometrium*. The broad ligaments pass from the uterus to the lateral walls of the pelvis. The *round ligaments* are situated between the folds of the broad ligament below the uterine tubes. The uterus normally lies in an anteverted position with the fundus facing the abdominal wall and the cervix towards the sacrum. There is also some anteflexion, the body being bent forwards on the cervix. The uterus is maintained in this position by the *transverse ligaments*, which run from the cervix to the lateral walls of the pelvis and by the *uterosacral ligaments*, which run from the cervix to the sacrum. The uterus is also supported indirectly by the *pelvic floor* (see Fig. 13.12).

If the ovum is *fertilized* in the uterine tube, it embeds in the thickened, vascular endometrium, which has been prepared to receive it by the action of the hormones. It remains there, increasing in size, until it fills the uterus after which the uterus grows with it until the end of the period of *pregnancy* (Fig. 26.8). At the site of implantation the *placenta* develops; this is the organ through which the fetus receives nourishment and oxygen from the maternal blood during intrauterine life.

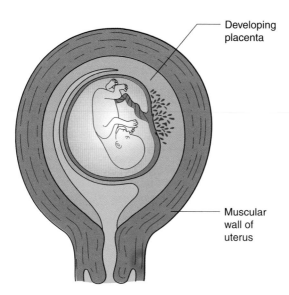

Developing
placenta

Muscular
wall of
uterus

Fig. 26.8 The uterus at about the tenth week of pregnancy, showing the placenta and developing fetus.

Menstruation is the shedding of the thickened endometrium, with some blood, which occurs each month after puberty until the menopause.

The anterior lobe of the pituitary secretes FSH, which initiates the development of a follicle in the ovary. As the *ovum* matures the follicle secretes oestrogen, which is necessary for the growth of the endometrium and its preparation to receive the fertilized ovum. Oestrogen is also responsible for the gradual development of secondary sex characteristics in the young girl. When the amount of oestrogen in the blood reaches a high level, further secretion of FSH is prevented but the anterior lobe of the pituitary begins to release *luteinizing hormone*. Following ovulation (the release of the ovum from the follicle), luteinizing hormone converts the ruptured follicle into the *corpus luteum*, which secretes the hormone progesterone. This hormone completes the development of the endometrium. If the ovum is not fertilized the corpus luteum begins to degenerate, the level of progesterone decreases and the endometrium is shed. The low progesterone level also stimulates the pituitary to secrete more FSH and the cycle begins again. For convenience of description the first day of the menstrual flow is designated Day 1 of the menstrual cycle. Ovulation usually occurs at about the 14th day (Fig. 26.9).

The *vagina* extends from the uterus to the labia; it lies behind the bladder and urethra and in front of the rectum and anal canal. The cervix enters the anterior wall of the vagina at right angles so the posterior wall of the vagina is longer than the anterior wall. The recesses formed by the projection of the cervix into the vagina are called fornices. The walls of the vagina are normally in contact with each other. The vagina has two coats: a muscular coat, which has longitudinal and circular fibres, and an inner lining of mucous membrane, which is in folds or rugae. After puberty, this lining becomes thick and is rich in glycogen; the action of certain bacteria (Döderlein's bacilli) on the glycogen makes the vaginal secretion slightly acidic. The peritoneum passes over the body of the uterus onto the back of the cervix and

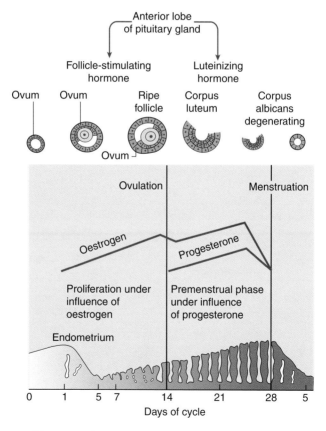

Fig. 26.9   Diagrammatic representation of the changes in the endometrium in the menstrual cycle.

close to the posterior fornix of the vagina before being reflected back in front of the rectum. This fold is called the recto-uterine pouch.

External organs

The external genital organs in the female are collectively called the vulva (Fig. 26.10) and consist of:

- The mons pubis
- The labia majora and minora
- The clitoris
- The vestibule of the vagina
- The greater vestibular glands.

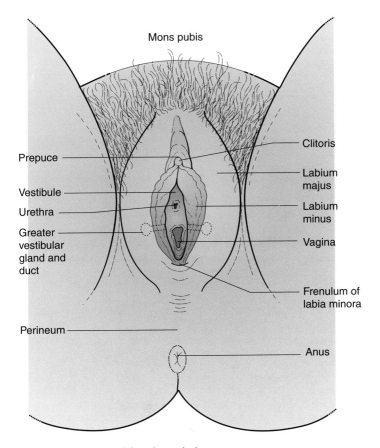

Fig. 26.10   The external female genital organs.

The *mons pubis* is a pad of fat covered with skin lying over the symphysis pubis. It bears hairs after puberty.

The *labia majora* are two folds of fatty tissue covered with skin extending backwards from the mons on either side of the vulva and disappearing into the perineum behind. They develop at puberty and are covered with hair on the outer surface after that stage. They atrophy after the menopause.

The *labia minora* are two smaller fleshy folds within the labia majora. They meet in front to form a hood-like structure called the *prepuce*, which surrounds and protects the *clitoris*. The labia minora unite behind in the frenulum of the labia minora (fourchette). This is merely a fold of skin, which is often torn in the first labour. The labia minora are covered with modified skin rich in sweat and sebaceous glands to lubricate their surfaces.

The *clitoris* is a small sensitive organ containing erectile tissue (corresponding to the male's penis). It lies in front of the vulva immediately below the mons pubis and is protected by the prepuce.

The *vestibule* is the cleft between the labia minora. The orifices of the vagina and urethra open into it.

The orifice of the urethra lies at the back of the vestibule, projecting slightly from the normal surface level. At the entrance, there are two fine tubular glands, called urethral glands, which secrete lubricating fluid and are more important because they tend to harbour infection in cases of gonorrhoea.

The orifice of the vagina occupies the space between the labia minora behind the vestibule. It is normally a slit from front to back, the side walls of the vagina being in contact. The orifice is largely blocked in the virgin by the *hymen*. This is a double fold of mucous membrane, usually crescent shaped, and leaving a gap at the front for the escape of the menstrual flow. Sometimes it has a number of small perforations; this is called a fenestrated hymen. Sometimes there is no opening but this is usually owing to the vagina not having developed into a canal. The hymen is placed a little inside the orifice of the vagina so that there is a slight depression between the frenulum of the labia and the hymen.

The greater vestibular glands are two small glands lying one under each labium majus on either side of the vaginal orifice. Their ducts open laterally to the hymen. They secrete a lubricating fluid to moisten the surface of the vulva and so facilitate sexual intercourse.

The whole surface of the vulva is covered with modified skin, i.e. stratified epithelium. It carries hair on the outer surfaces only although the inner surfaces are exceptionally rich in sebaceous and sweat glands so that the surfaces are moist and there is no friction when walking.

The *perineum* is the expanse of skin from the vaginal orifice back to the anus. It is about 5 cm in length and bears hair. It lies over the perineal body, a mass of muscle and fibrous tissue separating the vagina from the rectum. The muscle of the perineal body is largely the levator ani, the chief muscle of the pelvic floor.

SEXUAL INTERCOURSE

In order for normal fertilization to occur, whereby male and female gametes are brought together in the fallopian tubes, sexual intercourse has to take place. *Sexual intercourse* involves the insertion of the penis into the vagina and the subsequent deposition of seminal fluid in the vagina.

In the male, sexual excitement leads to erection of the penis. *Erection* takes place when the penis becomes engorged with blood, leading to enlargement in length and girth. After erection, it is possible for the male to insert the penis into the female's vagina. The sensations associated with sexual intercourse lead to intense sexual excitement and *orgasm* in the male, followed by *ejaculation* of seminal fluid into the vagina.

The female responds by erection of the clitoris, enlargement of the breasts and erection of the nipples. Preceding and during sexual intercourse the walls of the vagina release fluid, which provides lubrication. Orgasm may accompany sexual intercourse in females.

BREASTS

The *breasts* are two glands that secrete *milk* and are accessory organs to the female reproductive system. They are present in rudimentary form in the male. They lie on the front aspect of the thorax and vary considerably in size. They are circular in outline and convex anteriorly. In the centre of the surface is the nipple, which projects normally from the skin level and is pink in the virgin but pigmented after the first pregnancy (Fig. 26.11).

The breast is a compound saccular gland with its ducts converging to the nipple and opening on its surface in large numbers. The gland tissue is similar to the tissue of the sebaceous gland but more highly developed and secreting milk in place of sebum. The gland is divided into lobes by partitions of fibrous tissue – a fact that makes

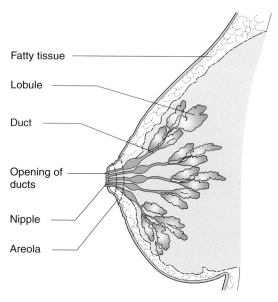

Fig. 26.11 The breast.

the draining of an abscess in the breast difficult. The lobes are sub-divided into lobules.

The breasts develop at puberty under the influence of hormones, and further development occurs during pregnancy as a result of hormones from the pituitary gland and ovaries. Fluid known as *colostrum* is secreted in small amounts by the gland during pregnancy and at the time of childbirth but the true milk is not secreted until about the third day of the puerperium.

SELF–TEST QUESTIONS

- Describe the structure of the testes.
- What is the function of the corpus luteum?
- What are the functions of oestrogen and progesterone in menstruation?
- What happens at fertilization?
- What is colostrum?

Bibliography

Kenworthy N, Snowley G, Gilling C 2002 Common foundation studies in nursing. Churchill Livingstone, Edinburgh

Kindlen S (ed) 2003 Physiology for health care and nursing, 2nd edn. Churchill Livingstone, Edinburgh

Martini FC 2004 Fundamentals of anatomy and physiology, 6th edn. Benjamin Cummings, San Francisco

Montague S, Watson R, Herbert R 2005 Physiology for nursing practice. Baillière Tindall, London

Watson R 1999 Essential science for nursing students. Baillière Tindall, London

Watson R, Fawcett TN 2003 Pathophysiology, homeostasis and nursing. Routledge, London

Index

Notes:
Page numbers in **bold** refer to tables
Page numbers in *italics* refer to figures

U

V